Wood Utilization

WOOD UTILIZATION

JEROME E. CARTER
EDITOR

Nova Science Publishers, Inc.
New York

Library of Congress Cataloging-in-Publication Data

Wood utilization / GAO.
p. cm.
Includes text from a GAO report to congressional requesters entitled: Wood utilization: federal research and product development activities, support, and technology transfer, June 2006.
Includes index.
ISBN 1-60021-459-2
ISBN 13: 978-1-60021-459-2
1. Lumbering--United States. I. United States. Government Accountability Office.
TS820.W66 2006
674--dc22 2006030378

Published by Nova Science Publishers, Inc. ✤ New York

CONTENTS

PREFACE

Wood utilization research and product development spans a broad spectrum of activities. These activities fall into five categories: harvesting, wood properties, manufacturing and processing, products and testing, and economics and marketing. This book deals with the U.S. federal input in this field.

In: Wood Utilization
Editor: J. E. Carter, pp. 1-46
ISBN: 1-60021-459-2

FEDERAL RESEARCH AND PRODUCT DEVELOPMENT ACTIVITIES, SUPPORT, AND TECHNOLOGY TRANSFER*

WHAT GAO FOUND

Federal wood utilization research and product development span a broad spectrum of activities. These activities fall into five categories: harvesting, wood properties, manufacturing and processing, products and testing, and economics and marketing. Of the 12 federal agencies that provided support to wood utilization research and product development, only the Forest Service and the CSREES-funded wood utilization centers had activities in all five categories; although all the agencies had activities in manufacturing and processing. Coordination of these activities is both informal and formal. Scientists informally coordinate their activities by conferring with each other and sharing information at conferences and professional meetings and through publications. In some cases, coordination occurs through more formal mechanisms, such as cooperative arrangements and other joint ventures.

During fiscal years 2004 and 2005, the 12 federal agencies made available at least $54 million annually for wood utilization research and product development activities, measured either in budget authority or expenditures. (Dollars are reported in either budget authority or expenditure data, depending on the availability of agency data.) The Forest Service made available about half of these funds. In addition, the Forest Service—the only agency that directly employs scientists and support staff to conduct wood utilization research and product

* Excerpted from GAO Report GAO 06-624, Dated June 2006.

development—reported having almost 175 full-time equivalent scientists and support staff in each of these years. For fiscal years 1995 through 2005, the Forest Service's budget authority for wood utilization research and product development activities fluctuated moderately from year-to-year (in inflation-adjusted dollars). In contrast, overall, CSREES' budget authority for the wood utilization research centers increased over the period (in inflation-adjusted dollars), in part because of the addition of four new wood utilization research centers between fiscal years 1999 and 2004.

To transfer technologies and products to industry, federal agencies generally rely on scientists and technology transfer specialists, who use methods such as information sharing, technical assistance, and demonstration projects. For example, applying research from the Forest Products Laboratory, Forest Service technology transfer specialists assisted a small forest products company in producing flooring from small trees by, among other things, providing solutions to product imperfections like warping and discoloration.

Abbreviations

CCA	chromated copper arsenate
CNC	computer-numeric controlled
CRADA	cooperative research and development agreement
CRIS	Current Research Information System
CSREES	Cooperative State Research, Education, and Extension Service
FTE	full-time equivalent
GPR	ground-penetrating radar
HUD	Department of Housing and Urban Development
OSB	oriented strandboard
RPA	The Forest and Rangeland Renewable Resources Planning Act of 1974
SBIR	Small Business Innovation Research
STTR	Small Business Technology Transfer
TMU	Technology Marketing Unit
USDA	U.S. Department of Agriculture
VOC	volatile organic compound

June 15, 2006
The Honorable Saxby Chambliss
Chairman, Committee on Agriculture, Nutrition and Forestry
United States Senate

The Honorable Richard G. Lugar
United States Senate

The Honorable Thad Cochran
United States Senate

More wood is consumed every year in the United States than all metals, plastics, and masonry cement combined. In residential construction, for instance, wood accounts for about 90 percent of structural framing and sheathing, as well as large portions of other wood products, including cabinets, doors, windows, moldings, millwork, sub-flooring, and finished flooring. Despite the nation's significant use of wood products, some segments of the U.S. forest products industry, such as furniture manufacturing, have declined. Over the past 10 years, forest products companies have consolidated, in part by closing or idling mills, to reduce costs and remain competitive with foreign companies.

The forest products industry is made up of a few large multinational companies and many medium and small companies.[1] To remain competitive, these companies have had to become more efficient and adapt to changing wood resources. For example, wood product companies had primarily used large trees, but these trees are now becoming scarce. At the same time, the supply of small-diameter trees has increased, in part because of federal, state, and local efforts to thin forests of these trees and therefore reduce the buildup of fuels that could contribute to large forest fires. To maximize their use of small-diameter trees and other wood resources, such as sawdust, companies rely on research into new methods of using wood. However, only a few large companies conduct wood utilization research and develop new products.

Federal research and product development in wood utilization helps provide the science and technology needed to conserve the nation's forest resources, supply the demand for wood products, and support forest management and restoration activities. At least 12 federal agencies support wood utilization research and product development. These include the U.S. Department of Agriculture's (USDA) Cooperative State Research, Education, and Extension Service (CSREES), Forest Service, and Natural Resources Conservation Service; the Department of Defense's (Defense) Army, Army Corps of Engineers, and

Office of Naval Research; the Department of Energy; the Department of Homeland Security's Coast Guard; the Department of Housing and Urban Development (HUD); the Department of the Interior's (Interior) Bureau of Indian Affairs; the National Science Foundation; and the Department of Transportation. However, historically, only two of these agencies—the Forest Service and CSREES—have had significant funds specifically targeted to wood utilization research and product development. The other agencies do not have defined programs for wood utilization research and product development but provided support through various mechanisms, such as grants.

Most of the Forest Service's wood utilization research and product development is conducted at its Forest Products Laboratory in Madison, Wisconsin, which is the agency's national laboratory for these activities; five of its regional research stations also conduct wood utilization research and product development. CSREES focuses on wood utilization research and product development through a grant directed by congressional committee (committee-directed grant) to 10 wood utilization research centers in 12 universities around the country.

Although many agencies have provided support for wood utilization research and product development, the Forest Service has the most experience in this area—it has been conducting wood utilization research and product development since 1910 and is a key player in carrying out these research and product development activities. However, a 2002 National Academy of Sciences report showed a nearly 30-percent decline in the Forest Service's budget authority for forest products research since 1980 (in inflation- adjusted dollars), and a loss of about 46 percent in Forest Service research scientists from 1985 to 1999.[2]

In this context, you asked us to review the status of federal agencies' support for wood utilization research and product development. For this report, we identified (1) the types of wood utilization research and product development activities supported by federal agencies and how these activities are coordinated; (2) the level of support federal agencies made available for these activities in fiscal years 2004 and 2005, and the changes in the level of support at the Forest Service and at the CSREESfunded wood utilization research centers from fiscal years 1995 through 2005; and (3) how the federal government transfers the technologies and products from its wood utilization research and product development activities to industry.

To collect and catalogue the types of wood utilization research and product development activities federal agencies conduct and the amount of financial and personnel support they provide for these activities, we collected and analyzed activity, budget authority, and expenditure data in agency databases and files. For purposes of our review, wood utilization research and product development refers to the activities that occur from harvesting wood through recycling wood and paper products. We collected financial data for fiscal year 2004 for each agency and for fiscal year 2005, if available. In some cases, these data were expenditure data, and in other cases, they were budget authority data.[3] In addition, we collected data for 11 fiscal years, 1995 through 2005, for the two agencies that historically have had funds dedicated to wood utilization research and product development—the Forest Service and CSREES' wood utilization research centers. We analyzed these data in both nominal dollars and dollars adjusted for inflation.[4] CSREES also provides funding for wood utilization research and product development as part of other research and development grant programs; we collected these data only for fiscal year 2004 because 2005 data were not available for these grant programs. We collected information on full-time equivalent (FTE) staff from the Forest Service for fiscal years 1995 through 2005; the Forest Service is the only agency that employs full-time scientists and support staff to conduct wood utilization research and product development. We reviewed the reliability of agencies' budget and expenditure data and determined that the data were sufficiently reliable for the purposes of this review. We also interviewed scientists, university researchers, technology transfer specialists, and industry officials around the nation to obtain information on wood utilization research and product development activities, financial and personnel support, and technology transfer efforts. In addition, we conducted site visits at a limited number of federal, university, and industrial facilities that focus on wood utilization research and product development—the Forest Products Laboratory; Forest Service facilities in Virginia, West Virginia, and Oregon; the wood utilization center at Oregon State University; and a Weyerhaeuser Company research laboratory in Washington State. Among other things, we asked these officials how federal agencies transfer technologies and products to industry and asked them for examples of successful federal technology transfer efforts. We performed our work between February 2005 and May 2006, in accordance with generally accepted government auditing standards. Appendix I provides a more detailed description of our scope and methodology.

RESULTS IN BRIEF

Federal wood utilization research and product development span a broad spectrum of activities, and coordination of these activities is both informal and formal. These activities fall into five categories: harvesting, wood properties, manufacturing and processing, products and testing, and economics and marketing. For example, with respect to manufacturing and processing, ongoing research in log scanning technology and equipment focuses on detecting knots and rot in a log in order to cut it for maximum use. Research in this area also examines the manufacturing of high-performance products from wood previously considered too small or unusable. Of the 12 federal agencies, only the Forest Service and the CSREES-funded wood utilization research centers had activities in all five categories; in contrast, all the agencies had activities in manufacturing and processing. Informally, scientists coordinate their activities by conferring with each other and sharing information at conferences and professional meetings and through publications. However, in some cases, more formal mechanisms have been established through legislative provisions, agency rulemaking, memorandums of understanding, and other joint ventures. For example, HUD has a partnership with the leaders of the home building, product manufacturing, insurance, and financial industries; and representatives of six federal agencies to develop technologies for improving the quality, durability, energy efficiency, and affordability of residential building materials, which includes wood.

During fiscal years 2004 and 2005, the 12 federal agencies made available at least $54 million annually for wood utilization research and product development, measured either in budget authority or expenditures. (Dollars are reported in either budget authority or expenditure data, depending on the availability of agency data.) The Forest Service made available about half of these funds. In addition, the Forest Service—the only agency that directly employs scientists and support staff to conduct this research—reported having almost 175 FTE scientists and support staff in each of these years. For fiscal years 1995 through 2005, the Forest Service received total budget authority of $268 million for wood utilization research and product development (or $289 million in 2004 inflation-adjusted dollars), while CSREES' budget authority was about $51 million (or $55 million in 2004 inflation-adjusted dollars). For fiscal years 1995 through 2005, the Forest Service's budget authority for wood utilization research and product development activities fluctuated moderately from year-to-year (in 2004 inflation-adjusted dollars). In contrast, overall, CSREES' budget authority for the wood utilization research centers increased over the period (in 2004 inflation-adjusted dollars), in part because of the addition of four new wood utilization research centers between fiscal years 1999 and 2004.

To transfer technologies and products to industry, federal agencies generally rely on scientists and technology transfer specialists, who use such methods as information sharing, technical assistance, and demonstration projects. In terms of providing technical assistance, for example, federal scientists helped a small company that produces flooring for the National Collegiate Athletic Association improve its manufacturing efficiency. In addition, the Forest Service has a dedicated technology transfer unit—the Technology Marketing Unit (TMU)—that has four technology transfer specialists with expertise in wood utilization and product development. These specialists work to improve the use of wood by transferring technologies developed by the Forest Service to industry and others. For example, applying Forest Products Laboratory research, these specialists assisted a small forest products company in producing flooring from small trees by, among other things, providing solutions to product imperfections like warping and discoloration.

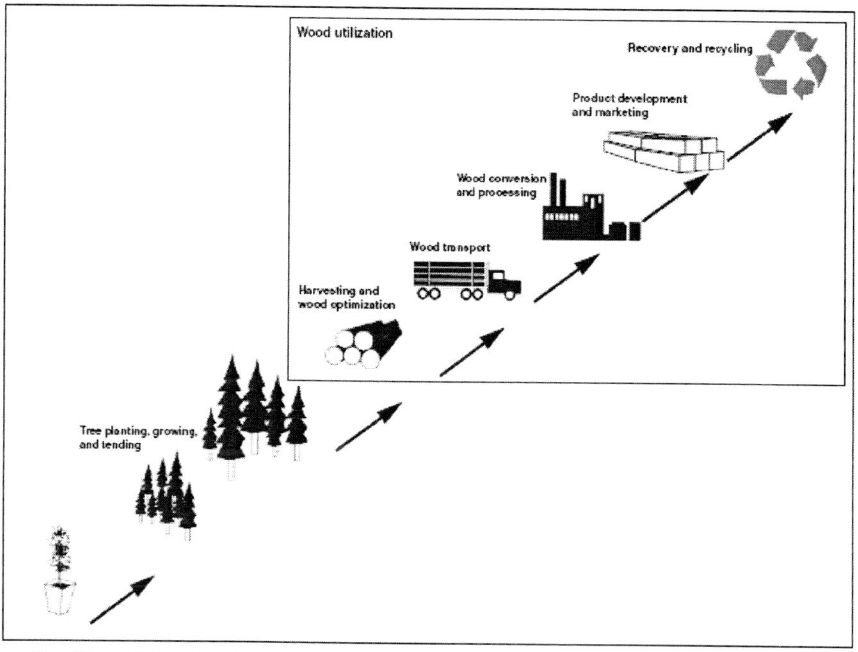

Source: Dr. Glenn Murphy, Oregon State University, 2005.

Figure 1. Wood Utilization Activities from Harvesting through Recycling.

BACKGROUND

The unique characteristics and relative abundance of wood have made it a natural material for a variety of uses, including homes and other structures, furniture, tools, vehicles, and decorative objects. Because wood varies in characteristics and volume by species, it may be heavy or light, stiff or flexible, and hard or soft. Federal agencies conduct research on the range of processes that occur between the time a tree is grown in the forest to the time it becomes a wood product and then is recycled. For purposes of our review, wood utilization research and product development refers to the activities that occur from harvesting the wood through the recycling of wood and paper products. (See fig. 1.)

THE FOREST PRODUCTS INDUSTRY

According to the North American Industry Classification System,[5] the U.S. forest products industry is divided into two sectors: wood product manufacturing and pulp and paper manufacturing. The wood product manufacturing sector comprises small companies, while the pulp and paper manufacturing sector tends to have fewer, larger companies.

The wood product manufacturing sector can be broken into three sub-sectors: (1) primary producers—sawmills and plywood mills; (2) secondary producers—millwork, cabinet, and furniture manufacturers; and (3) structural and reconstituted products producers—oriented strandboard (OSB), I-Joist, laminated veneer lumber, medium density fiberboard, and particleboard. The United States is the world's leading producer of lumber and wood products used in residential construction and in commercial wood products. According to 2004 data (the most recent data available), the wood product sector employed 535,000 workers nationwide and produced shipments valued at $103 billion.[6]

The pulp and paper manufacturing sector includes two industry groups: (1) manufacturers of pulp and paper and (2) manufacturers of products made from purchased paper and other materials, such as paper bags or tissues. The vast majority of the raw material for making paper is the residue from other mills—primarily chips from sawmills. The United States is also a leader in the pulp and paper business, producing about 28 percent of the world's pulp and 25 percent of the total world output of paper and paperboard. In 2004 (the most recent data available), the paper manufacturing sector employed 440,000 workers nationwide and produced shipments valued at $154 billion.

According to a federal government report, the U.S. forest products industry faces increasing competition from its traditional competitors (Canada, the Scandinavian countries, and Japan), as well as from emerging competitors (Brazil, Chile, and Indonesia).[7] Domestic purchases of paper and paperboard declined from 2000 to 2002, but have begun to rebound since then.[8] Approximately

120,000 jobs were lost in the paper manufacturing sector from 1999 to 2004, representing a 21.5-percent loss. Sectors of the wood product manufacturing industry have also declined. According to a 2003 Forest Service report, during the last decade, the wood household furniture industry lost approximately one-third of its market share to imports. China now accounts for one-third of U.S. imports, up from none a decade ago.[9]

Federal Agencies Support Wood Utilization Research and Product Development

Federal research and product development in wood utilization helps provide the science and technology needed to conserve the nation's forest resources, supply the demand for wood products, and support forest management and restoration activities. At least 12 federal agencies support wood utilization research and product development activities, but only 2 of these agencies—the Forest Service and CSREES—have programs targeted for these activities.

Primary Agencies—Forest Service and CSREES' Wood Utilization Research Centers

For the Forest Service, the Forest and Rangeland Renewable Resources Act of 1978 is the primary legislation authorizing the Secretary of Agriculture to implement a comprehensive research program for forest and rangeland renewable resources, including wood utilization, and to disseminate the results.[10] Other relevant legislation includes the following:

- The Biomass Research and Development Act of 2000, which requires the secretaries of Agriculture and of Energy to cooperate on policies and procedures that promote research and development leading to the production of fuels and biobased products; the act also established the Biomass Research and Development Initiative.[11]
- The Energy Policy Act of 2005 established technical areas for focusing research under the Biomass Research and Development Initiative.[12]
- The Healthy Forests Restoration Act of 2003 established a grant program to encourage the commercialization of woody biomass.[13]

The Forest Service's research and development organization establishes research work units in the field by developing formal mission statements, which must be approved by the Deputy Chief for Research and other senior managers.[14] A team from the Deputy Chief's Office and station directors' office formally reviews these mission statements and the unit's work at least every 5 years, and the review often includes input from the public and private sectors.

The Forest Service's wood utilization research and product development is carried out by scientists and professional support staff in 27 research work units around the country that were operating at the time of our review. Most of the Forest Service's wood utilization research and product development takes place at 16 research work units in the Forest Products Laboratory, which conducts research of national and international scope. The other 11 research work units are located in the Forest Service's Northeastern, Southern, Pacific Northwest, Pacific Southwest, and Rocky Mountain Research Stations, and these units mostly focus on regional wood utilization issues.[15] For example, research work unit 4104 of the Southern station focuses on managing Southern pine ecosystems, whereas research work unit 4701 of the Northeastern station focuses on efficiently using northern forest resources. These research work units produce 5-year research work plans that identify the mission, the problem to be solved through research, the proposed research approach, planned accomplishments, and staffing needs.

CSREES provides support for wood utilization research and product development through several grant programs. CSREES awards committee-directed grants to 10 designated wood utilization research centers at 12 universities. The first three centers were established in fiscal year 1985 at Oregon State University, Mississippi State University, and Michigan State University. These three centers were established to support wood utilization and harvesting research on western conifers, southern pine, and eastern hardwoods, respectively. In fiscal year 1993, three centers with specific research focuses were added at the University of Maine, the University of Minnesota at Duluth, and North Carolina State University. In fiscal year 1999, the University of Tennessee and the Inland Northwest Forest Products Research Consortium were added. The consortium consists of the universities of Idaho and of Montana, and Washington State University. The most recent additions are the University of Alaska Southeast, in fiscal year 2000, and West Virginia University, in fiscal year 2004. Every year each center submits a grant proposal, reviewed by CSREES staff, containing information on proposed research activities, budgets, and progress to date. Funding supports scientists and graduate students and helps to pay for new equipment, supplies, and travel.

In addition, CSREES provides grants to state-supported colleges and universities that can be used for, but are not specifically focused on, wood utilization research and product development through the following:

- The McIntyre–Stennis Act, a formula grant program, for forestry research, including two of eight potential funding areas focused on wood utilization and product development.[16]
- The Hatch Act, a formula grant program, designed to fund a number of broad agricultural research areas.[17]
- The National Research Initiative,[18] a competitive grant program with several research areas, including biobased products and energy. Wood utilization research and product development grants have been awarded under this initiative, as well as under CSREES' Small Business Innovation and Research grants and other small grants programs.

Other Agencies That Support Wood Utilization Research and Product Development

Ten other agencies also support wood utilization research and product development. Table 1 provides information on these agencies' principal authorizing legislation and a description of the programs that have supported wood utilization research and product development, and the mechanisms used for program delivery.

Technology Transfer

The Federal Laboratory Consortium for Technology Transfer defines technology transfer as "the process by which existing knowledge, facilities or capabilities developed under federal research and development funding are utilized to fulfill public and private needs."[19] Since 1978, Congress has enacted a series of laws to promote technology transfer and to provide technology transfer mechanisms and incentives. Table 2 presents selected laws that support technology transfer for wood utilization research and product development.

Table 1. Federal Agencies That Support Wood Utilization Research and Product Development, Principal Authorizing Legislation, and Description of Selected Programs

Federal agency	Principal authorizing legislation[a]	Program description
USDA—National Resource Conservation Service[b]	Biomass Research Development Act of 2000, Title III of the Agricultural Risk Protection Act of 2000 (Pub. L. No. 106-224)	Administers and funds grants for the Biomass Research and Development Initiative, under which competitively awarded grants, contracts, and financial assistance are provided to, or entered into with, eligible entities to carry out research on—and development and demonstration of—biobased fuels and biobased products (including woody biomass), and the methods, practices, and technologies, for their production.
Defense—Army, Corps of Engineers, Office of Naval Research	2005 Defense Appropriations Act	Provides committee-directed grants or contracts to specific universities to conduct wood utilization research and product development.
Department of Energy	Energy Policy Act of 1992 (Pub. L. No. 102-486), Biomass Research Development Act of 2000 (Pub. L. No. 106-224, tit.III (2000)	Enters into cost-share cooperative agreements and contracts with its national laboratories, private industry, and universities to conduct research on energy-efficient processes in energy-intensive industries, including the pulp, paper, and wood products manufacturing industries. Also develops technology for converting biomass into energy and chemicals.
Department of Homeland Security—Coast Guard	2002 Department of Transportation Appropriations Act (Pub. L. No. 107-87)	Provides a committee-directed contract to conduct wood utilization research and product development at a specific university.
HUD	Housing and Urban Development Act of 1970 (Pub. L. No. 91-609)	Administers the Partnership for Advancing Technology in Housing program, an interagency partnership that provides grants and financial assistance for research on residential housing materials, which includes wood.
Interior—Bureau of Indian Affairs	Snyder Act of 1921	Can award grants to support wood product development.
Department of Transportation	Section 1039 of the Intermodal Surface Transportation Efficiency Act of 1991 (Pub. L. No. 102-240, § 1039, 23 U.S.C. 144 nt); Section 401 of the Pipeline Safety Act of 1992 (49 U.S.C 112), Transportation Equity Act for the 21st Century (Pub. L. No. 105-178)	The Federal Highway Administration has awarded grants for research to increase the usage of timber and wood products in highway bridges, including improving the design and performance of timber structures and developing engineering design criteria for structural wood products for use in highway bridges.[c]

Table 1. (Continued)

Federal agency	Principal authorizing legislation[a]	Program description
Department of Transportation (Continued)		The Research and Innovative Technology Administration established centers of excellence and provides grants to advance technology and expertise in all areas of transportation, including transportation structures made from wood.
National Science Foundation	National Science Foundation Act of 1950 (Pub. L. No. 81-507)	Funds basic research at universities, small businesses, and other organizations. Wood utilization research could be funded under the National Science Foundation's broad research categories of engineering, chemistry, biology, social science, and education. Does not target wood utilization research and does not fund product development.

Sources: Legislation and agency documents.

[a] All laws cited are as amended.

[b] In 2006, USDA's Rural Development Agency assumed responsibility for this grant program from the Natural Resources Conservation Service.

[c] There was no new funding for this program in fiscal year 2004, according to Department of Transportation officials.

In addition to these laws, Executive Order 12591 ("Facilitating Access to Science and Technology") directs federal agencies to encourage and facilitate collaboration among federal laboratories, state and local governments, universities, and the private sector—particularly small business—in order to assist in the transfer of technology to the marketplace.

Technology transfer is also carried out through the nation's extension system, established by the Smith-Lever Act in 1914,[20] to assist in the development of practical applications of research knowledge in agriculture, including wood utilization. Under this system, thousands of county and regional extension specialists bring university expertise to the local level. Funding is provided by CSREES through annual formula grants to supplement state and county funds for extension services. The funds can be used for natural resources, including forestry or wood utilization, depending upon the priorities of the university.

The Renewable Resources Extension Act of 1978 created the Renewable Resources Extension Program.[21] Under this program, CSREES provides funds to 72 universities, which use these funds, along with state, local, and institutional funds, to deliver educational programs to forest and rangeland owners and managers. The program also provides guidance to states in developing their general extension programs for, among other things, timber utilization, harvesting, and marketing; wood utilization; and wood products marketing. These efforts have included wood utilization extension services, usually through extension specialists.

Table 2. Selected Laws That Support Technology Transfer for Wood Utilization Research and Product Development

Laws supporting technology transfer[a]	Description
General technology transfer laws	
The Stevenson-Wydler Technology Innovation Act of 1980 (Pub. L. No. 96-480)	Enacted to stimulate improved utilization of federally funded technology developments—including inventions, software, and training technologies—by state and local governments and the private sector.
The Patent and Trademark Law Amendments Act of 1980 (Bayh-Dole Act, Pub. L. No. 96-517)	Allowed universities, not-for-profit corporations, and small businesses to patent and commercialize their federally funded inventions.
The Federal Technology Transfer Act of 1986 (Pub. L. No. 99-502)	Authorized federal agencies to permit the directors of government-owned laboratories to enter into cooperative research and development agreements (CRADA) and to negotiate licensing agreements for inventions created in the laboratories.
The Technology Transfer Commercialization Act of 2000 (Pub. L. No. 106-404)	Broadened CRADA licensing to authorize federal laboratories to grant licenses to federally owned inventions for which a patent application was filed before the CRADA was signed.
Laws for research and development conducted by small businesses, which can include wood products companies	
Small Business Innovation Development Act of 1982 (Pub. L. No. 97-219)	Established the Small Business Innovation Research (SBIR) program, requiring federal agencies to reserve a portion of their research and development effort for awards to small businesses.
Small Business Research and Development Enhancement Act of 1992 (Pub. L. No. 102-564)	Extended the SBIR program, increased the percentage of an agency's budget to be devoted to SBIR, and established the Small Business Technology Transfer Program (STTR), a program under which a portion of a federal agency's extramural research or research and development effort is reserved for awards to small businesses.
Small Business Reauthorization Act of 2000, enacted as part of the Consolidated Appropriations Act for Fiscal Year 2001(Pub. L. No. 106-554) Laws focused on natural resources, including wood and its uses	Directed the Small Business Administration and participating agencies to, among other things, expand the scope of publicly available information on specific grants, and to annually report on their SBIR programs.

Table 2. (Continued)

Laws supporting technology transfer[a]	Description
Cooperative Forestry Assistance Act of 1978 (Pub. L. No. 95-313)	Authorized the Forest Service to carry out a program of technology implementation to ensure that new technology is introduced and forest resources research findings are made available to state forestry personnel, private landowners, wood processors, forest operators, and others.
The National Forest-Dependent Rural Communities Economic Diversification Act of 1990, enacted as part of the Food, Agriculture, Conservation and Trade Act of 1990 (Pub. L. No. 101-624)	Created the Economic Action Program to upgrade existing industries to use natural resources (including wood) more efficiently, and to expand the economic base of rural communities to alleviate or reduce their dependence on National Forest System land resources.

Source: GAO analysis of laws and regulations.
[a]All laws are as amended.

Federal Wood Utilization Research and Product Development Activities Fall into Five Categories and Are Coordinated Both Informally and Formally

Wood utilization research and product development conducted by 12 federal agencies span a broad spectrum of activities, and coordination of these activities is both formal and informal. These activities fall into five broad categories: (1) harvesting, (2) wood properties, (3) manufacturing and processing, (4) products and testing, and (5) economics and marketing.

Federal Wood Utilization Research and Product Development Activities Can Be Grouped into Five Broad Categories

We grouped the wood utilization research and product development activities that the 12 agencies conduct into five broad categories: harvesting, wood properties, manufacturing and processing, products and testing, and economics and marketing. Table 3 shows the definitions we used for the five categories and provides examples of the types of the research and product development activities that fall into each of these categories.

Table 3. Categories of Major Wood Utilization Research and Product Development Activities

Category	Definition	Examples of research and development activities
Harvesting	Using scientific and engineering principles to ensure cost-effective, environmentally acceptable, and safe forest operations, including planning, road building, harvesting, handling and processing, and transportation	• constructing forest roads • providing safety training • developing equipment to reduce soil compaction • using equipment to remove trees at the stump
Wood properties	Studying the basic and applied physical, chemical, and mechanical properties of wood and wood fiber to determine the suitability of this material for various uses, from pulp to structural beams to recycled composite products	• examining the basic surface properties of different wood and wood-based materials to determine interaction with contaminants in water for improved water repellency • examining reactions to mold, mildew, fungi, and various temperatures • research on reactions to moisture • research on the suitability of dead or dying trees for use in several new and emerging wood-processing technologies
Manufacturing and processing	New and better manufacturing ways to extract, reduce, and convert virgin wood raw materials to useful products and the development of technology to allow the re-use of materials and products to the maximum extent possible	• examining the process for removing moisture from wood and the impact of the drying processes on the strength, durability, and color of the wood • studying and developing log scanning technology and equipment to detect knots and rot • examining the manufacturing of high-performance products from wood previously considered too small, unsuitable, or defective, and from recycled wood • using ultrasound to detect internal defects in wood • improving paper manufacturing technologies to reduce energy consumption and improve paper quality • improving bonding of particles, flakes or fibers, and adhesives in the manufacturing of composite wood products, such as oriented strandboard (OSB)[a]

Table 3. (Continued)

Category	Definition	Examples of research and development activities
Products and testing	Developing test methods and gathering and evaluating data on the differing uses of wood and wood fiber products	• testing chemical and other treatments designed to prevent the deterioration or destruction of wood or to extend its service life • studying the long-term durability, safety, and structural performance of adhesively bonded wood assemblies and the development of better testing methodologies for the durability of wood products, such as composite siding • testing to develop a performance-based moisture design approach for wood frame buildings that includes interior moisture design, exterior moisture design, and performance of wood products under various moisture and temperature regimes • developing analytical computer programs and other methods to assess the potential moisture accumulation in building components
Economics and marketing	Evaluating and tracking (1) domestic and international supply and demand trends, trade policies, and markets, including market opportunities; and (2) harvesting and production costs for alternative material and energy inputs and processing options	• describing the cost and price effects of changes in forest management practices and forest sector policies • analyzing factors affecting the near- and long-term outlook for supply and demand • analyzing broad-scale trends in trade and investment and their effects on forest products industries and firms • analyzing marketing and distribution patterns in international trade for furniture, lumber, and other raw materials • analyzing the stages of the manufacturing process to reduce the cost of finished wood products

Sources: GAO's analysis of agencies' activities and consultation with Forest Service officials.

[a]OSB is a composite wood product made of layered wood strands, often used as an alternative to plywood.

Table 4 shows the types of research and product development activities and examples of these activities by agency. All 12 agencies had activities in the manufacturing and processing category.

Table 4. Agencies' Wood Utilization Research and Product Development Activities and Examples of These Activities

Federal agency	Research and product development categories of activities	Examples of research and product development activities
USDA		
Forest Service	• Harvesting Wood properties • Manufacturing and processing • Products and testing • Economics and marketing	• examine and improve a variety of different types of machinery for tree removal and processing and high-technology computers to measure trees as they are processed • understand factors that lead to poor wood surface quality and their influences on wood bonding • optimize how wood, woody biomass, and recycled and nonwood materials are converted into durable cost-effective, high-performing, and long-service-life products • conduct fundamental research in the areas of solid wood products, composites, and paper manufacture • evaluate the effects of technology trends and market changes on forest management
CSREES	• Harvesting • Wood properties • Manufacturing and processing • Products and testing • Economics and marketing	• new equipment that reduces soil compaction and increases timber production • wood preservative technologies that result in longer product life • improved machining technologies to increase the speed of manufacturing wood products at lesser cost • automated lumber grading and machining systems using optical scanners, machine vision, and radio frequency fields to detect and cut around defects in lumber • recycling processes for converting newsprint into composite board
Natural Resources Conservation Service[a]	• Manufacturing and processing	• adding value to woody biomass by processing material into lumber and poles • using woody biomass as fuel for electricity generation • using biomass for wood-burning facilities to cogenerate power and steam
Federal agency	Research and product development categories of activities	Examples of research and product development activities

Table 4. (Continued)

Defense		
Army	• Wood properties Manufacturing and processing	• studying the use of composite materials to construct modular ballistic protective shelters, which include the development of ballistic panels with a wood layer in the panel design
Army Corps of Engineers	• Wood properties • Manufacturing and processing	• examining the use of wood composite materials in constructing temporary facilities and quarters that are lightweight, affordable, rapidly erectable, modular, protective, and blast or ballistic resistant
Office of Naval Research	• Wood properties • Manufacturing and processing	• developing low-cost and stronger wood composite materials to replace Navy wood structures, such as pier components (e.g., decking and fender components)
Department of Energy	• Manufacturing and processing	• reducing the energy intensity of manufacturing processes in the pulp and paper and wood products industries • studying drying technologies to reduce the energy required to remove water from the pulp used to make paper
Department of Homeland Security—Coast Guard	• Wood properties • Manufacturing and processing	• designing, building, and demonstrating a pier made of wood composite
HUD	• Wood properties • Manufacturing and processing	• study of insulated composite wood panels in residential construction
Interior—Bureau of Indian Affairs	• Manufacturing and processing	• studying the feasibility of developing wood products using woody biomass • studying the feasibility of using woody biomass to generate electricity and heat greenhouses
Department of Transportation	• Wood properties • Manufacturing and processing	• studying preservatives and coatings for structural wood products for highway bridges • studying wood composite materials for transportation
National Science Foundation	• Wood properties • Manufacturing and processing • Products and testing	• studying the mixing of plastics with wood fibers to create wood-composite • products that are used in doors, windows, decks siding, and roofs • improving the conversion of wood chips and other biomass to paper fibers

Source: GAO analysis of agency documents.

[a] In 2006, USDA's Rural Development Agency assumed responsibility for this research effort from the Natural Resources Conservation Service.

The Forest Service and CSREES were the only two agencies that had wood utilization research and product development activities in all five categories. According to our analysis of the Forest Service's 27 research work units' plans covering fiscal years 1995 through 2005, over 80 percent of wood utilization research and product development occurred in three categories: wood properties, products and testing, and manufacturing and processing. In addition, CSREES wood utilization research centers' annual research proposals for the same period showed that about 70 percent of their activities occurred in the following three categories: wood properties, manufacturing and processing, and economics and marketing. According to a CSREES official, the CSREES wood utilization research centers are allowed by law to use the funding to conduct technology transfer activities, which are reflected in the economics and marketing category.

Appendixes II and III, respectively, provide detailed information on wood utilization research and product development activities for the Forest Service, for multiyear periods (beginning in the late 1980s) to the present; and CSREES, for fiscal years 1995 through 2005.

Federal Wood Utilization Research and Product Development Activities Are Coordinated Both Informally and Formally

We found instances of both informal and formal coordination of federal activities for wood utilization and product development. According to many scientists at the Forest Service, informal coordination occurs among the relatively small wood utilization research and product development community of scientists, and these scientists are often aware of related scientific research. Scientists share information at scientific and industry conferences and professional meetings and through publications, and in some cases work informally to share staff and equipment. Specific examples include the following:

- One Forest Service scientist associated with the Southern Research Station—with 30 years of experience in wood utilization research on Douglas Fir—shares resources and expertise with the Pacific Northwest Research Station on the plantation growth of this species.
- Forest Service scientists in the Southern Research Station have collaborated with colleagues in Australia, Denmark, Japan, and New Zealand on using wood from southern forests to develop wood composite products. These collaborative efforts were established primarily through professional relationships.

- A Forest Service scientist at the Pacific Northwest Research Station told us that scientists use annual professional meetings, such as those held by the Forest Products Society and the Society of Wood Science and Technology, as important mechanisms for coordinating their work and broadening the scope of their research area.

The CSREES wood utilization research centers reported that they have more informal than formal coordination mechanisms with other wood utilization research centers and federal agencies. Like the Forest Service, these informal mechanisms include sharing information with their colleagues through professional meetings, publications, and newsletters.

We also identified some formal mechanisms to coordinate wood utilization research and product development that are set up through legislative provisions, agency rulemaking, memorandums of understanding, cooperative arrangements, and other joint ventures. Specific examples include the following:

- The Biomass Research and Development Act of 2000 requires USDA and the Department of Energy to carry out a Biomass Research and Development Initiative under which competitively awarded grants, contracts, and financial assistance are provided to eligible entities to carry out research on fuels and products derived from biomass, including woody biomass. The agencies work together on developing grant solicitations, reviewing grant proposals, and selecting recipients. The act also created a Biomass Research and Development Board, co-chaired by the Department of Energy and USDA, to coordinate programs within the federal government for promoting the use of biobased fuels and products. The board's mission is to maximize the benefits from federal grants and assistance by promoting collaboration and avoiding duplication of effort through strategic planning on biomass research. The board has approved the formation of a federal Woody Biomass Working Group to coordinate and focus federal efforts on woody biomass utilization.
- For 40 years, Forest Service wood utilization scientists have had standing annual meetings with representatives from both the paper and pulp and solid wood industries to present research results and obtain input and review from industry. When updating their research work unit plans every 5 years, these scientists also seek advice from outside sources, including industry representatives, academics, and environmental groups.
- Scientists also participate in research consortiums or cooperative arrangements with industry. For example, scientists in the Forest

Service's Southern Research Station participate in a consortium studying wood quality that has members from nine companies, including Weyerhaeuser and Georgia Pacific. CSREES wood utilization research centers also form cooperative arrangements. According to an Oregon scientist, these research cooperatives typically consist of 10 to 12 partners. The cooperatives set a research agenda and formally coordinate research through annual meetings and reports; each university, as well as government agencies, are asked to contribute funding annually. For example, scientists at the University of Minnesota wood utilization research center formed a productivity cooperative that includes state, county, university, and industry members (such as International Paper) to continue to strengthen applied forestry concepts and ensure the sustainability of Minnesota's forest products industry.

- The Forest Service's Northeastern Research Station formed the Furniture Steering Committee, which is composed of furniture manufacturers, consultants, equipment manufacturers, state economic development agencies, and universities to provide guidance on furniture research programs at the station and elsewhere. The steering committee recommended research on more efficient manufacturing and "just-in-time" training, which has been integrated into the research work unit's plan.
- HUD's Partnership for Advancing Technology in Housing is a voluntary partnership between leaders of the home building, product manufacturing, insurance, and financial industries; and representatives of six federal agencies concerned with housing.[22] These six agencies work with HUD to develop technologies to improve the quality, durability, energy efficiency, and affordability of residential building materials; these materials could include wood. For example, with the partnership's support, the Forest Service's wood chemistry research work unit has been able to work cooperatively with laboratories in Japan, Sweden, and Finland on developing coatings to protect wood from the effects of weathering.
- Forest Service scientists at the Southern Research Station's Utilization of Southern Forest Resources work unit have a memorandum of understanding with the Chinese government to host post-doctoral students from China; the station has hosted 25 students in the past 5 years. These students serve as additional staff resources to help the research work unit carry out its research activities.

- To construct a forest biomass life cycle assessment model, several partners established a joint venture: the Forest Service's Pacific Southwest Research Station; the California Energy Commission's Public Interest Energy Research Program; the University of California at Davis; several state and federal agencies; and energy, forestry, and environmental consultants. Partners will use the model to identify and analyze the social, economic, and environmental costs and benefits of using forest biomass to generate electrical power. This research project is planned in three phases over a 3- to 5-year period. Each participant shares in the cost of the venture.

Federal Agencies Made Available at Least $54 Million Annually for Wood Utilization Research and Product Development in Fiscal Years 2004 and 2005; Forest Service Support Fluctuated Moderately, and CSREES Support Increased Over 10 Years

The 12 federal agencies we reviewed made available at least $54 million annually in financial support for wood utilization research and product development activities in fiscal years 2004 and 2005, measured either in budget authority or expenditures.[23] Furthermore, the Forest Service employed almost 175 scientists and support staff in each of these two fiscal years. From fiscal years 1995 through 2005, the Forest Service received total budget authority of $268 million for wood utilization research and product development (or $289 million in 2004 inflation-adjusted dollars) while CSREES' budget authority for the wood utilization research centers was about $51 million (or $55 million in 2004 inflation-adjusted dollars). For fiscal years 1995 through 2005, the Forest Service's budget authority for wood utilization research and product development activities fluctuated moderately from year-to-year (in 2004 inflation-adjusted dollars). Over the same period, overall, CSREES' budget authority for the wood utilization research centers increased (in 2004 inflation-adjusted dollars), in part because four new wood utilization research centers were added during fiscal years 1999, 2000, and 2004.

The Forest Service Provided Most of the Support for Wood Utilization Research and Product Development in Fiscal Years 2004 and 2005

The 12 federal agencies we identified as supporting wood utilization research and product development made available at least $54.4 million in financial support for this work, measured in either budget authority or expenditures,[24] in fiscal year 2004, the year with the most complete data available. For fiscal year 2005, the agencies made available at least $54.3 million. Our data for fiscal year 2005 are complete except for data for the CSREES grants funded under the McIntyre-Stennis Act and the Hatch Act; the National Research Initiative; Small Business Innovation Research grants; and other small grants. See table 5.

As table 5 shows, the Forest Service made available about half of the financial support for conducting wood utilization research and product development. In fiscal year 2004, the Forest Service made available about 52 percent of the $54.4 million, while four other agencies—CSREES, the Department of Energy, the National Science Foundation, and the Natural Resources Conservation Service—made available about 44 percent of the support; the remaining seven agencies together made available about 5 percent of the $54.4 million.

Of the $54.4 million made available in fiscal year 2004, about $34 million ($28.3 million for the Forest Service and $5.7 million for the CSREES wood utilization research centers) was directly targeted to wood utilization research and product development. In addition, $1.9 million of other support targeted for wood utilization research and product development was made available by the Army, the Coast Guard, and the Office of Naval Research through committee-directed funding to specific universities to conduct research on wood composites.

The remaining $18.5 million of the $54.4 million was made available in fiscal year 2004 from grant programs not targeted to wood utilization research and product development. That is, wood utilization research and product development was not the sole purpose of the grant or program. The Department of Energy made available the largest amount of this nontargeted support—$7.4 million. CSREES provided $3.0 million in fiscal year 2004 to support other wood utilization research and product development through grant programs authorized under the McIntyre-Stennis Act and the Hatch Act; the National Research Initiative; Small Business Innovation Research grants; and other small grants. The Natural Resources Conservation Service made available grant funding to promote greater innovation and development in all forms of biomass—including agricultural and woody biomass—with $5.3 million targeted to woody biomass research, under the Biomass Research Development Act of 2000. The other agencies made available the remaining $2.8 million.

Table 5. Federal Financial Support in Wood Utilization Research and Product Development, by Agency, Fiscal Years 2004-2005

Dollars in thousands		
Department	**2004 financial support[a]**	**2005 financial support[a]**
USDA		
Forest Service	$28,251[b]	$27,179[b]
CSREES	8,710[c,d]	5,820[b,e]
Natural Resources Conservation Service	5,269[b]	4,627[b]
Defense		
Army Research	25[b]	1,050[b]
Army Corps of Engineers	0	2,395[b]
Office of Naval Research	1,459[f]	1,424[f]
Department of Energy	7,419[b,g]	6,233[b,g]
Department of Homeland Security—Coast Guard	442[f]	351[f]
HUD	0	225[b]
Interior—Bureau of Indian Affairs	486[b]	276[b]
Department of Transportation	63[b]	441[b]
National Science Foundation	2,270[f]	4,242[f]
Total	**$54,394**	**$54,263**

Sources: Agency documents, CSREES' Current Research Information System, and National Science Foundation's Project Reports Summary and Search and Awards databases.

[a] Financial data are presented in either budget authority or expenditures, as indicated.

[b] Budget authority.

[c] Includes both budget authority and expenditures.

[d] Includes $5.67 million for wood utilization research centers and $3.04 million for the other CSREES grants funded under the McIntyre-Stennis Act and the Hatch Act; the National Research Initiative; Small Business Innovation Research Grants; and other small grants.

[e] Data for other CSREES grants were not available for 2005.

[f] Expenditures.

[g] Budget authority for the Industrial Technologies Program.

Of the 12 agencies, only the Forest Service directly employs full-time scientists and support staff to conduct wood utilization research and product development. Most of these employees work at the Forest Products Laboratory, as shown in table 6.

Table 6. FTE Staff for Wood Utilization Research and Product Development at the Forest Service, Fiscal Years 2004-2005

	FTE scientists		FTE support staff		Total FTE staff	
Forest Service unit	**(2004)**	**(2005)**	**(2004)**	**(2005)**	**(2004)**	**(2005)**
Forest Products Laboratory	62.8	59.6	58.3	57.3	121.1	116.9
Northeastern Research Station	9	9	11	10	20	19
Pacific Northwest Research Station	9	9.3	4.5	7	13.5	16.3
Pacific Southwest Research Station	1	1	0	2	1	3
Rocky Mountain Research Station	0.1	0.1	0	0	0.1	0.1
Southern Research Station	8.7	8.7	9.3	9.8	18	18.5
Total	90.6	87.7	83.1	86.1	173.7	173.8

Source: Forest Service documents.

The other 11 agencies we reviewed do not have full-time federal scientists dedicated to wood utilization research and product development, and were unable to provide information on scientists and support staff working on federal wood utilization research and product development activities.

From Fiscal Years 1995 Through 2005, Forest Service Budget Authority for Wood Utilization Research and Product Development Fluctuated Moderately from Year-To-Year

For fiscal years 1995 through 2005, the Forest Service received total budget authority for wood utilization research and product development of $268 million (which is equivalent to $289 million in 2004 inflation-adjusted dollars). As table 7 shows, during this 11-year period, the annual budget authority ranged between $24.2 million and $28.2 million (in 2004 inflation-adjusted dollars), with moderate fluctuations from year-to-year.

Table 8 shows the total FTE scientists and support staff for the Forest Service's wood utilization research work units, from fiscal years 1995 through 2005.

Table 7. Forest Service Budget Authority for Wood Utilization Research and Product Development, Fiscal Years 1995-2005

Dollars in thousands												
	1995	**1996**	**1997**	**1998**	**1999**	**2000**	**2001**	**2002**	**2003**	**2004**	**2005**	**Total**
Nominal dollars	$23,727	$20,873	$21,416	$21,616	$22,196	$23,195	$26,041	$26,726	$27,246	$28,251	$27,179	**$268,465**
2004 dollars	$28,037	$24,201	$24,404	$24,336	$24,666	$25,264	$27,711	$27,907	$27,899	$28,251	$26,451	**$289,128**

Source: Forest Service documents.

Note: Totals may not add due to rounding.

Table 8. Forest Service FTE Staff for Wood Utilization Research and Product Development, Fiscal Years 1995-2005

	1995	**1996**	**1997**	**1998**	**1999**	**2000**	**2001**	**2002**	**2003**	**2004**	**2005**
Scientists	98.2	98.3	90.6	85.0	82.8	84.2	86.6	84.1	86.1	90.6	87.7
Support staff	109.0	101.5	88.5	87.4	86.4	87.8	89.8	89.3	88.3	83.1	86.1
Total	**207.2**	**199.8**	**179.1**	**172.4**	**169.2**	**172.0**	**176.4**	**173.4**	**174.4**	**173.7**	**173.8**

Source: Forest Service documents.

Table 9. The Forest Products Laboratory's Budget Authority for Wood Utilization Research and Product Development, Fiscal Years 1995-2005

Dollars in thousands												
	1995	1996	1997	1998	1999	2000	2001	2002	2003	2004	2005	Total
Nominal dollars	$17,596	$15,800	$15,800	$15,726	$15,850	$15,850	$17,924	$18,551	$19,088	$20,025	$19,213	$191,423
2004 dollars	$20,792	$18,319	$18,004	$17,705	$17,614	$17,265	$19,074	$19,370	$19,545	$20,025	$18,698	$206,411

Source: Forest Service documents.

As figure 2 shows, over the period, the levels of budget authority (adjusted for inflation) and FTE staff for wood utilization research and product development at the Forest Service fluctuated moderately. From fiscal year 1995 to fiscal year 1996, both budget authority (in 2004 inflation-adjusted dollars) and FTE staff at the Forest Service decreased by 14 percent and 4 percent, respectively. After 1996, budget authority for the most part increased through 2004 and then decreased in 2005.

FTE staff continued to decrease through 1999, increased in 2000, and thereafter remained relatively stable. (See app. IV for information on changes in FTE Forest Service scientists and support staff for wood utilization research work units for each year from fiscal year 1995 through 2005.)

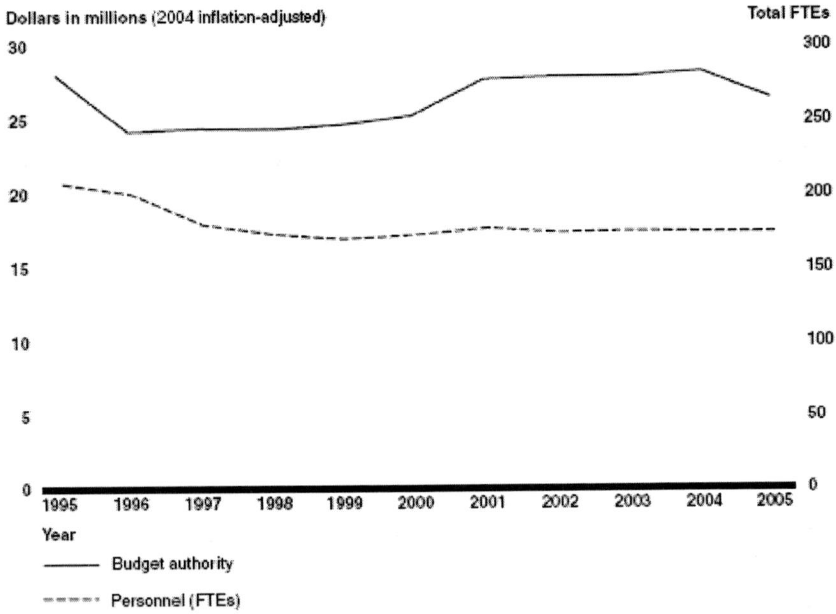

Source: GAO analysis of Forest Service documents.

Figure 2. Total Budget Authority for Forest Service Wood Utilization Research and Product Development, and FTE Staff, Fiscal Years 1995-2005.

During the 11-year period, the Forest Products Laboratory's budget authority also fluctuated moderately. Between fiscal years 1995 and 2000, the budget authority declined by 17 percent (in 2004 inflation-adjusted dollars), from $20.8 million to $17.3 million; it increased again from fiscal years 2001 through 2004, but was still lower in 2005 than in 1995. (See table 9.)

Table 10 shows the total FTE scientists and support staff for the Forest Products Laboratory's wood utilization research work units, from fiscal years 1995 through 2005.

Table 10. The Forest Products Laboratory's FTE Staff for Wood Utilization Research and Product Development, Fiscal Years 1995-2005

	1995	1996	1997	1998	1999	2000	2001	2002	2003	2004	2005
Scientists	67.2	69.0	64.3	60.6	56.4	55.8	59.0	58.5	57.2	62.8	59.6
Support Staff	78.0	73.5	61.5	60.4	61.9	60.3	62.3	61.5	62.0	58.3	57.3
Total	145.2	142.5	125.8	121.0	118.3	116.1	121.3	120.0	119.2	121.1	116.9

Source: Forest Products Laboratory.

The number of FTE Forest Products Laboratory scientists and support staff generally declined from fiscal years 1995 through 2000; then it fluctuated moderately. Figure 3 shows the changes in budget authority and FTE scientists and support staff at the Forest Products Laboratory. See appendix IV for funding and FTE staff, by research work unit, at the Forest Products Laboratory and at the research stations for fiscal years 1995 through 2005.

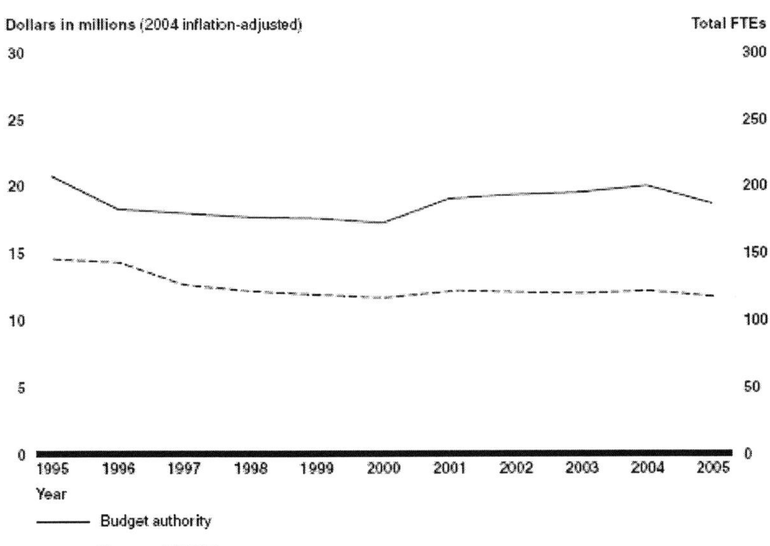

Source: GAO analysis of Forest Service documents.

Figure 3. Total Budget Authority for the Forest Products Laboratory's Wood Utilization Research and Product Development, and FTE Staff, Fiscal Years 1995-2005.

While financial support for wood utilization research and product development at the Forest Service has fluctuated moderately during the past 11 years, Forest Service scientists and managers expressed concerns about resource constraints. They noted that increases in budget authority cover salary increases and other fixed costs, but that these increases may not be enough to cover increases in the costs of other operating expenses—such as purchasing or calibrating equipment, obtaining laboratory supplies, and traveling for research. The Forest Products Laboratory's operating budget authority declined by about 67 percent between fiscal years 1995 and 1998 (in 2004 inflation-adjusted dollars), from about $1.95 million to $650,000; it also fluctuated within a narrow range from fiscal years 1999 to 2005, ending with $630,000. (See table 11.)

Table 11. The Forest Products Laboratory's Operating Budget, Fiscal Years 1995-2005

Dollars in thousands												
	1995	1996	1997	1998	1999	2000	2001	2002	2003	2004	2005	Total
Nominal dollars	$1,651	$1,337	$ 901	$577	$630	$613	$672	$642	$636	$647	$647	$8,953
2004 dollars	$1,951	$1,550	$1,027	$650	$700	$668	$715	$670	$651	$647	$630	$9,859

Source: Forest Service documents.

Figure 4 shows changes in the dollars available for operating expenses (adjusted to 2004 dollars) in fiscal years 1995 through 2005 at the Forest Products Laboratory.

Many of the scientists with whom we spoke cited instances in which fewer resources had diminished their ability to conduct research. For example, according to one scientist, he is spending less time in the laboratory because he is devoting more time to obtaining outside funding for his research work unit. Another scientist told us that his research work unit must now limit the number of wood samples from private sources that the unit has time to analyze, which it did not need to do in the past. According to Forest Service officials, due in part to funding constraints, as well as to better serve the scientific community, the Forest Products Laboratory has developed a strategic plan, and is in the process of reorganizing and consolidating its research work units and reducing the number of scientists and support staff.[25]

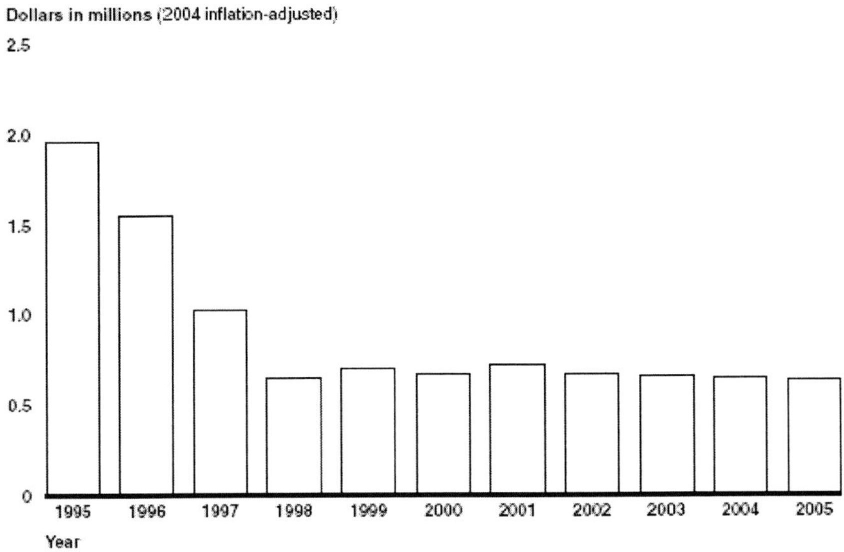

Source: GAO analysis of data from Forest Service documents.

Figure 4. The Forest Products Laboratory's Operating Budget, Fiscal Years 1995-2005.

From Fiscal Years 1995 Through 2005, CSREES' Wood Utilization Research Centers' Budget Authority and the Number of Centers Increased

Table 12 shows that the total budget authority for fiscal years 1995 through 2005 for CSREES' wood utilization research centers was about $51.2 million (which is equivalent to $54.8 million in 2004 inflation-adjusted dollars), and figure 5 illustrates that, overall, CSREES' budget authority (adjusted for inflation) for the wood utilization research centers increased over the period. The increase in budget authority was due in part to the addition of four new wood utilization research centers, particularly when two new centers were added in fiscal year 1999; new centers were added again in fiscal years 2000 and 2004.

Table 12. CSREES Budget Authority for Wood Utilization Research Centers, Fiscal Years 1995-2005

Dollars in thousands											
	1995	1996	1997	1998	1999	2000	2001	2002	2003	2004	2005	Total
Nominal dollars	$3,530	$3,532	$3,305	$3,305	$4,805	$4,805	$5,400	$5,304	$5,730	$5,670	$5,820	$51,206
2004 dollars	$4,172	$4,095	$3,766	$3,721	$5,340	$5,234	$5,747	$5,538	$5,867	$5,670	$5,664	$54,814

Source: CSREES documents.

Table 13. Total Budget Authority for CSREES Wood Utilization Research Centers, Fiscal Years 1995-2005

Dollars in thousands (2004 inflation-adjusted)											
	1995	1996	1997	1998	1999	2000	2001	2002	2003	2004	2005
Wood utilization research centers											
Maine	$889	$872	$802	$793	$916	$797	$877	$843	$827	$736	$698
Michigan	889	873	802	793	916	797	877	843	827	736	698
Minnesota	275	270	249	246	284	247	272	261	252	228	216
Mississippi	889	872	802	793	916	797	877	843	1,290[a]	1,154[a]	1,148[a]
North Carolina	342	336	309	305	353	307	338	325	313	283	269
Oregon	889	872	802	793	916	797	877	843	814	736	698
Consortium[b]					572	497	547	526	523	460	496
Tennessee					468	407	448	430	422	376	406
Alaska						588	634	622	599	543	586
West Virginia										418	451
Total	$4,172	$4,095	$3,766	$3,721	$5,340	$5,234	$5,747	$5,538	$5,867	$5,670	$5,664

Source: CSREES documents.

[a] The large increase in Mississippi grants in fiscal years 2003 through 2005 occurred because of an additional committee-directed grant for a special statewide forest resources inventory.

[b] The Inland Northwest Forest Products Research Consortium consists of the universities of Idaho and of Montana, and Washington State University.

Note: Totals may not add due to rounding.

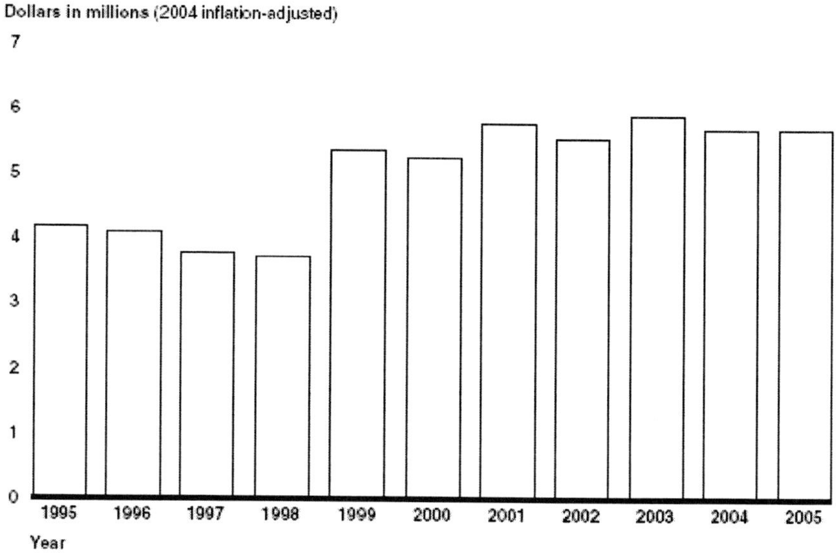

Source: GAO analysis of data from CSREES documents.

Figure 5. Total Budget Authority for CSREES Wood Utilization Research Centers, Fiscal Years 1995-2005.

While the increase in the number of wood utilization research centers would suggest an increased commitment to wood utilization research and product development, after adjusting for inflation, most of the centers, individually, experienced a downward trend in budget authority, as table 13 shows. (See app. IV for wood utilization research centers' budget authority in nominal dollars over the period.)

Federal Agencies Rely on Scientists and Specialists to Transfer Technology through a Variety of Methods

The 12 federal agencies generally rely on scientists and technology transfer specialists to transfer technologies to industry through a variety of methods,[26] such as information dissemination, technical assistance, demonstration projects, and patents and licensing. While federal scientists are involved in some technology transfer, their primary responsibility is research; in contrast, specialists are responsible solely for technology transfer. In addition, the Forest Service has a unit dedicated to transferring the results of wood utilization research and product development: the Forest Service's TMU. We identified a number of examples of activities that have occurred using each of the technology transfer methods, mostly from the Forest Service and CSREES wood utilization research centers.

Scientists and Technology Transfer Specialists Disseminate Information

Scientists are expected to transfer the results of their work and primarily disseminate information through publications—particularly those in peer-reviewed journals—which help establish the validity of their research results. The Forest Service counts the number of articles published in these journals to assess scientists' performance and reports this information as a performance measure for research in its annual report to Congress.[27] Furthermore, according to Forest Service scientists, some industry officials may also read and use these journals. For example, a window and door manufacturer used the information from a journal article on the characteristics of wood from smaller trees for use in composites to develop a new and higher-value use for this wood. Instead of burning the wood as waste, the manufacturer now uses it in his products.

Scientists also disseminate research results to industry through a variety of other methods, including publications that are not peer reviewed, Web sites, presentations of their work at professional meetings, and workshops. Specific examples include the following:

Publications that are not peer reviewed include the Forest Service's one-page information sheets, *TechLines*; technical reports; industry magazines; trade journals; and training manuals. For example, one training manual was developed after industry representatives asked a Forest Service scientist to create a publication on avoiding accidents caused by improperly constructed logging trails. Scientists also contribute to user manuals that are important to the building industry and homeowners, such as *Finishes for Exterior Wood*—20,000 copies sold in the past 10 years; and the *Wood Mold Maintenance Manual*—10,000 copies in circulation.

Most of the Forest Service's wood utilization research work units maintain Web sites that list articles or provide links to articles and contact information. For example, a research work unit in the Southern Research Station reported that 18,335 distinct users—approximately 1,528 per month—accessed its Web site in 2004, downloading 37,376 publications. Some of CSREES' wood utilization centers also have Web sites, and some scientists have their own Web sites devoted to their wood utilization research and product development.

The Forest Service's State and Private Forestry's Wood Education and Resource Center in West Virginia offers a grant program to transfer research results. In one instance, grant funds helped support the issuance of three newsletters informing pallet producers, shippers, and technical assistance personnel of the latest developments in implementing new international regulations. These regulations require that all pallets crossing international boundaries be treated to prevent the spread of invasive species. Additionally, three technical bulletins summarizing the results of the center's applied research in this area were developed and distributed to an international audience.

- Workshops conducted by scientists for industry include the University of Minnesota's industry-specific training on streamlined manufacturing procedures to over 75 companies, which has resulted in partnerships with 15 of them. University of Minnesota scientists reported that these partnerships have led to productivity improvements of 50 to 75 percent and cost reductions of 25 to 50 percent, with estimated financial impacts of over $750,000.
- Forest Service scientists have shared information through broadcasts. A radio host in Arkansas has a weekly show on forestry issues, and scientists from the Southern Research Station have discussed their research.
- The Forest Products Laboratory conducts "Entrepreneur Tours" in which small- to medium-size mill operators from western states tour the Forest Products Laboratory to learn about current research and how they can use it.

Technology transfer specialists—at the Forest Service's State and Private Forestry program and extension specialists and programs at universities—also play a key role in disseminating information to industry. As of February 2006, the Forest Service employed nine technology transfer specialists, who also provide other types of assistance to small businesses. Like scientists, specialists reach industry and other users through Web sites and publications—particularly those that are not peer reviewed, like trade journals, newsletters, and industry magazines. Specialists sometimes work directly with scientists to disseminate research to targeted users. For example, technology transfer specialists at Louisiana State University's extension program publish the *Dry Kiln Club* newsletter, which provides updated research results from the university's scientists on wood-drying and moisture-related wood decay to an audience of over 1,000.

Extension specialists also disseminate information through targeted group education to industry and other users. This education includes short courses, continuing education courses, and workshops. Specialists often develop these courses using the results of research conducted at their university and other universities, the Forest Service, and other federal and state agencies. Specific examples include the following:

- Extension specialists at Virginia Tech University offered 27 short courses to industry in calendar year 2004. In one of these courses, they combined research from the College of Business with their own knowledge of wood science to teach methods for selling wood products.
- Extension specialists in Ohio taught a multiweek course to landowners on how to prune and manage their trees and market their products. The course was designed to help the landowners take advantage of a new pallet plant soon to be opening in their area.
- Extension specialists at Mississippi State's wood utilization research center have provided logger education to over 3,000 logger firms during the past 10 years.
- Extension specialists at West Virginia University's Appalachian Hardwood Center have conducted technology transfer and outreach efforts for the past 15 years. For example, in October 2004, the center hosted a log-sawing and grading workshop that focused on the efficient grading and recovery of lumber for low-grade logs.
- To enhance competitiveness in the region's forest products industry, the University of Tennessee's Forest Products Center has a wood products extension specialist who conducts workshops, issues newsletters, and takes other actions to transfer information from the CSREES wood utilization research center to industry.

Technology transfer specialists also attend industry and professional conferences and meetings, where they present information and meet with industry representatives to build their networks. In addition, they disseminate information by creating directories that provide contact information for wood industries in their state.

Technical Assistance Is an Important Tool for Transferring Technology to Industry

Both scientists and technology transfer specialists provide technical assistance through (1) telephone calls; (2) hands-on technical assistance; and (3) software development.

Both scientists and technology transfer specialists respond to telephone calls requesting assistance from industry, consumers, and homeowners.

For example, one scientist at Oregon State University estimated receiving over 200 calls per year; another scientist estimated receiving over 400. Forest Products Laboratory managers estimated that they receive 4,000 such calls per year.

Scientists and technology transfer specialists also provide industry and others with hands-on technical assistance. Examples include the following:

- Forest Products Laboratory scientists provided technical assistance to help a small company improve its manufacturing efficiency by applying research on the fasteners and connectors used to assemble and disassemble portable flooring. This company produces flooring for the National Collegiate Athletic Association.
- Forest Products Laboratory scientists helped a large drumstick manufacturer solve a durability problem by developing a way to inject drumsticks with a polymer to strengthen them.
- Forest Products Laboratory scientists provide technical assistance by identifying wood samples for companies, as well as for private citizens. As part of this wood identification, they assist manufacturers in resolving problems they have in using different types of woods with different finishes. In 2004, they identified 600 specimens for industry, 350 specimens for government agencies, and 370 specimens of wood for the general public.
- For 12 years, the University of Minnesota has worked with a company to provide support in material selection, prototyping, performance testing, and market assessment and development. These efforts have helped the company introduce several new product lines in office furniture, store fixtures, and cabinet components; expand from 30 to 450 employees; and increase the company's sales from $5 million to $50 million annually over the period.

- The Department of Energy offers energy assessments of facilities that manufacture wood products or produce pulp and paper, although the department requires a substantial cost investment from the company. According to the Department of Energy, these assessments have resulted in an annual savings of up to $9 million for some companies.

Agencies also develop software and make it available, often for free, on Web sites. For example, a Forest Service computer program developed by researchers at the Forest Service's Northeastern Research Station provides a realistic simulation model that allows industry to identify more efficient strategies to reduce waste in the manufacturing process. More than 700 computer program packages have been sent to industry, and follow-up telephone calls by Forest Service scientists indicate that the program is being used in planning and optimization activities by many of the recipients. Similarly, the Department of Energy's Industrial Technologies Program provides free software tools to the forest products industry to improve energy efficiency in industry processes.

Demonstration Projects Can Highlight the Application of Wood Utilization Research and Product Development

Agencies also transfer research results through demonstration or pilot projects in mills, plants, and on-site at research locations. Specific examples include the following:

- The Forest Products Laboratory built a research demonstration house in 2001 on-site. The research in the demonstration house focuses on improving the use of traditional wood products, recycled and engineered wood composites, natural disaster resistance, energy efficiency, and indoor air quality. Features include a permanent wood foundation and engineered wood composites in the roof.
- In cooperation with the homebuilding and forest products industries, the Forest Products Laboratory constructed a house on the Washington, D.C., mall as part of the 2005 annual Smithsonian festival. The house showcases new technologies developed by the Forest Products Laboratory and cooperators, such as manufacturers of structural insulated panels. The house was visited by several thousand people over the course of the 10-day festival.

- Forest Products Laboratory scientists helped a company implement a demonstration project in its saw mill. The project showed that, with improvements to the company's machinery for determining lumber quality, the company could increase efficiency by as much as 12 percent—thus adding an estimated $1.2 million annually in profit.
- Scientists at West Virginia University's wood utilization research center have developed a new technology for using oak as a raw material in the manufacture of OSB. The Weyerhaeuser Company and other industry partners are testing the process and the produced strands in test runs to verify the results. If successful, the research work unit anticipates lower raw material costs and increased use of oak as an engineered wood product component. Success could lead to new or expanded OSB manufacturing facilities, and new jobs, in the Appalachian region.
- Forest Service scientists at the Southwest Wildland/Urban Interface and Forest Health Restoration research work unit, in Flagstaff, Arizona, have joined with Northern Arizona University on framing techniques using small-diameter logs. This partnership has led to a demonstration project with the Navajo Nation to develop hogans using small-diameter wood. Hogans are traditional housing structures tribes still use, and are typically built with more costly wood from larger trees.
- HUD, through its Partnership for Advancing Technology in Housing program, helped a builder in North Carolina to demonstrate the durability and cost of various building materials (including insulated composite wood panels) in four residential duplex units. The builder agreed to build each duplex out of a different building material, and HUD is evaluating the materials' performance at this site.
- The Office of Naval Research has several demonstration projects in place using wood-plastic composite materials to replace wooden pier components, such as deck boards and fendering components. Such demonstrations help Navy engineers become familiar with new technologies and their benefits before the technologies are widely available.
- The Coast Guard, in a contract with the University of Maine for composite wood research, requires the university to demonstrate that the composite structures it developed could be used in a marine environment and be more durable than traditional structures. The university will build a dock for the Coast Guard to demonstrate the use of the composite material it has developed.

Results of Wood Utilization Research and Product Development Can Be Patented and Then Licensed to Industry

Technology can also be transferred to industry through licensing and patenting. The Forest Service employs one full-time patent attorney, stationed at the Forest Products Laboratory, to assist scientists in patenting inventions they create as part of federally sponsored research projects; industry can then license these patents.[28] The Forest Service Patent and Licensing Program handles all aspects of patents and licensing, including reviewing invention disclosures, filing and prosecuting patent applications, negotiating patent licenses and other technology transfer-related agreements. Between January 1, 1995, and December 3, 2005, a total of 58 patents were issued, and 12 applications related to wood utilization are currently pending, according to the Forest Service.

Technology can also be transferred to industry through licensing and patenting. The Forest Service employs one full-time patent attorney, stationed at the Forest Products Laboratory, to assist scientists in patenting inventions they create as part of federally sponsored research projects; industry can then license these patents.[28] The Forest Service Patent and Licensing Program handles all aspects of patents and licensing, including reviewing invention disclosures, filing and prosecuting patent applications, negotiating patent licenses and other technology transfer-related agreements. Between January 1, 1995, and December 3, 2005, a total of 58 patents were issued, and 12 applications related to wood utilization are currently pending, according to the Forest Service.

Scientists at the CSREES wood utilization centers also obtain patents on processes and products they have developed. For example, scientists at the University of Minnesota's wood utilization research center have obtained over 20 patents that they have then licensed to private industry. These patents include those for extracting chemicals from birch bark that can be used in medicine, in manufacturing absorbent panels, and in a foam-and-wood composite log used for siding. They also reported having a number of pending patent applications in the areas of housing systems and the extraction of natural chemicals from birch bark waste products.

The Forest Service Has a Unit Dedicated to Transferring Wood Utilization Research and Product Development

The Forest Service has a unit dedicated to transferring the results of wood utilization research and product development activities—the TMU, part of the

State and Private Forestry Program, located at the Forest Products Laboratory. TMU's mission is to improve wood utilization by transferring technologies developed primarily by the Forest Products Laboratory and other Forest Service research units. As of February 2006, TMU employed four technology transfer specialists with expertise in wood utilization and product development. These specialists collaborate with Forest Service scientists, primarily at the Forest Products Laboratory, to provide technical assistance to local governments, private landowners, rural communities, and forest industries to ensure the ready adoption of technologies based on forest materials.

Like scientists and other technology transfer specialists, TMU's specialists disseminate research results through publications, conferences, and workshops. Specific examples include the following:

- In fiscal years 2004 and 2005, TMU reported distributing 40,000 and 6,900 publications, respectively.[29] For example, TMU's newsletter, the *Forest Products Conservation and Recycling Review*, has a circulation of over 800. In fiscal year 2005, it published 19 issues of *TechLines* on topics ranging from the outdoor performance of wood-plastic composites; to wood flooring and roofing; to using waste wood for filtering water.
- TMU participated in 45 workshops, conferences, presentations, training sessions, and exhibits in fiscal year 2004 that were attended, in total, by over 5,000.
- In 2004, TMU cosponsored the SmallWood conference in Sacramento, California, that was attended by over 350, including harvesting contractors, rural development officials, community leaders, forest products business owners, environmental groups, and tribes.
- TMU provided an updated software tool that allows users to compare the unit costs of various heating fuels—the Fuel Value Calculator—allowing wood to be compared to conventional fossil fuels, such as natural gas or fuel oil. The calculator is available on TMU's Web site.[30]

In addition, since TMU's technology transfer specialists are located on-site with Forest Products Laboratory scientists, they have an opportunity to learn about the research from its early stages. Furthermore, when a technology is developed, the specialists can work with the scientists to conduct a market analysis to determine potential applications. For example, in 2004, TMU published *Assessing the Market Potential of Roundwood Recreational Buildings*,[31] which provides information on the applicability of the Forest Products Laboratory's research on roundwood.

TMU also transfers technology to users by providing technical assistance directly to industry, communities, and individuals nationwide, as well conducting demonstration projects. Specifically, TMU specialists perform the following activities:

- *Answer numerous phone inquiries and letters, and host visitors—over 2,000 in both fiscal years 2004 and 2005.* Specialists provide answers to technical questions, point a user to key information sources, or provide a link and contact information to researchers working in a user's area of interest.
- *Travel to facilities to provide hands-on advice and answer questions.* For example, TMU assisted a remote California logging community hard-hit by mill closures to create over 100 new jobs through a small forest products company and a nonprofit training center. Applying Forest Products Laboratory research, TMU specialists helped the company specialize in producing flooring from small-diameter trees by, among other things, providing solutions to product imperfections like warping and discoloration.
- *Work with companies and communities in implementing research results or new technology through pilot and demonstration projects.* For example, TMU staff are working with the Department of Energy's National Renewable Energy Laboratory on a project testing small-scale biomass modular units, called "BioMax 15s," that use wood chips to create electricity. The technology is still in the pre-commercial phase, so the department and the TMU are using a demonstration program at several sites around the country, including a high school in Walden, Colorado, and a furniture-making business at the Zuni Pueblo in New Mexico.

In addition to its technology transfer responsibilities, in fiscal year 2005, the unit led the evaluation of proposals for USDA's Woody Biomass Grant Program. This program made available over $4 million in grants designed to increase the utilization of woody biomass from or near National Forest System lands. The program is designed to improve forest restoration activities by using and creating markets for small-diameter material and low-valued trees that were removed during activities to reduce hazardous fuels. Grants could range in value from $50,000 to $250,000.

AGENCY COMMENTS

We provided a draft of this report for review and comment to USDA's CSREES, Forest Service, and Natural Resources and Conservation Service; Defense; Department of Energy; Department of Homeland Security; HUD; Interior; Department of Transportation; and the National Science Foundation. The Forest Service, DOT, Energy, and Interior provided technical comments, which we incorporated as appropriate. CSREES, Natural Resources and Conservation Service, Defense, Department of Homeland Security, HUD, and the National Science Foundation did not have comments on the draft report.

As agreed with your offices, unless you publicly announce the contents of this report earlier, we plan no further distribution until 7 days after the date of this letter. At that time, we will send copies of this report to interested congressional committees; the Secretaries of Agriculture, Defense, Energy, Homeland Security, Housing and Urban Development, Interior, and Transportation; the Director of the National Science Foundation; the Director of the Office of Management and Budget; and other interested parties. We will also make copies available to others upon request. In addition, the report will be available at no charge on the GAO Web site at *http://www.gao.gov.*

If you have any questions about this report, please contact me at (202) 512-3841 or *nazzaror@gao.gov.* Contact points for our Offices of Congressional Relations and of Public Affairs may be found on the last page of this report. GAO staff who made major contributors to this report are listed in appendix V.

Robin M. Nazzaro
Director, Natural Resources and Environment

REFERENCES

[1] Environmental Protection Agency, *Profile of the Pulp and Paper Industry, 2nd edition,* EPA/310-R-02-002 (Washington, D.C.: Nov. 2002), and Environmental Protection Agency, *Profile of the Lumber and Wood Products Industry*, EPA/ 310-R-95-006 (Washington, D.C.: Sept. 1995).

[2] National Research Council, *National Capacity in Forestry Research* (Washington, D.C.: National Academy of Sciences, 2002).

[3] Dollars are reported in either budget authority or expenditure data, depending on the availability of agency data. Most agencies and programs

allocate a portion of their budget authority for wood utilization, in part in response to direction contained in appropriations committee reports. Those budget authority amounts are reported when available. However, the only data available for some CSREES programs and for the National Science Foundation were expenditure data.

[4] We adjusted nominal dollars using the Department of Commerce's Fiscal Year Chain-Weighted Price Index for the Gross Domestic Product with 2004 as the base year.

[5] A business classification system, adopted in 1997, developed by the U.S., Mexican, and Canadian governments to provide comparable statistics across the three countries. It classifies business into sectors, subsectors, and industry groups.

[6] U.S. Census Bureau, *Statistics for Industry Groups and Industries, Annual Survey;* MO4(AS)-1 (Washington, D.C.: Dec. 2005).

[7] U.S. Department of Energy, *Forest Products Industry of the Future, Fiscal Year 2004 Annual Report* (Washington, D.C.: Feb. 2005).

[8] Ince, Peter J.; Akim, Edward; Lombard, Bernard; and Parik, Tomas; Chapter 8, "Higher demand and production in 2004, but growth wavers in 2005: Markets for paper, paperboard and woodpulp, 2004-2005. United Nations Economic Commission for Europe, *Forest Products Annual Market Review, 2004-2005*. Timber Bulletin Vol. LVIII (2005).

[9] Schuler, Albert; and Buehlmann, Urs; *Identifying Future Competitive Business Strategies for the U.S. Residential Wood Furniture Industry: Benchmarking and Paradigm Shifts*, U.S. Forest Service, Northeastern Research Station, report NE-304.

[10] Pub. L. No. 95-307 (1978) (as amended).

[11] Pub. L. No. 106-224, tit.III (2000).

[12] Pub. L. No. 109-58, § 941 (e)(2), 119 Stat. 875-76 (2005).

[13] Pub. L. No. 108-148, § 203, 117 Stat. 1901 (2003).

[14] These include the headquarters office's topic area staff director, and research stations' directors and assistant directors.

[15] The Northeastern Research Station is located in Newtown Square, Pennsylvania; Pacific Northwest, in Portland, Oregon; Pacific Southwest, in Albany, California; Rocky Mountain, in Fort Collins, Colorado; and Southern, in Asheville, North Carolina.

[16] Formula grants are grants distributed to state and local governments using formulas that are based on data such as state population and personal income. Under the McIntyre–Stennis Act, CSREES apportions funds among participating states, considering factors such as nonfederal expenditures for

forestry research by state-certified eligible institutions, areas of nonfederal commercial forest land, and the volume of timber cut annually. States, in turn, determine the proportionate amounts of assistance to be extended to these qualified state-supported institutions.

[17] Act of March 2, 1887, ch. 314, 24 Stat. 440 (as amended).

[18] E.g., Pub. L. No. 89-106, § 2, Aug. 4, 1965, 79 Stat. 431 (1965) (as amended).

[19] The Federal Laboratory Consortium for Technology Transfer is a nationwide network of federal laboratories that provide the forum for developing strategies and opportunities that link technology with laboratory missions and the marketplace. It was organized in 1974 and formally chartered by the Federal Technology Transfer Act of 1986 to promote and to strengthen technology transfer nationwide.

[20] Act of May 8, 1914.

[21] Pub. L. No. 95-306 (1978).

[22] The six agencies are the Department of Energy, the Environmental Protection Agency, the Federal Emergency Management Agency, the Department of Commerce's National Institute of Standards and Technology, the National Science Foundation, and USDA. In fiscal years 2004 and 2005, the Environmental Protection Agency, the Federal Emergency Management Agency, and the National Institute of Standards and Technology did not support wood utilization research and product development.

[23] See footnote 3.

[24] See footnote 3.

[25] USDA, *Strategic Framework of Forest Products and Utilization Research and Development (FPURD),* (Washington, D.C.: Feb. 2006)

[26] For the purposes of this report, the term "technology transfer specialist" includes marketing and utilization specialists and university extension specialists.

[27] *Forest Service Performance and Accountability Report*, Fiscal Year 2004, p. 208

[28] Since 1980, the Bayh-Dole Act and subsequent executive actions have generally given federal contractors, grantees, and cooperative agreement funding recipients the option to retain ownership rights to, and profit from, commercializing the inventions they create as part of federally sponsored research projects. In return for these rights, these recipients are required to file for patent protection, pursue commercialization of the inventions, give preferences to small businesses in licensing, ensure that any products

resulting from the inventions are substantially manufactured in the United States, and comply with certain reporting requirements.

[29] The higher number in 2004 is due to distribution of 30,000 copies of a special publication, "Better Rural Places."

[30] http://www.fpl.fs.fed.us/tmu/

[31] Paun, Dorothy; Randall Cantrell, and Susan LeVan-Green. Forest Products Lab, FPL-GTR-144.

In: Wood Utilization
Editor: J. E. Carter, pp. 47-52
ISBN: 1-60021-459-2

APPENDIX I: OBJECTIVES, SCOPE, AND METHODOLOGY

This report describes (1) the types of wood utilization research and product development activities supported by federal agencies and how these efforts are coordinated; (2) the level of support federal agencies made available for these activities in fiscal years 2004 and 2005, and changes in the level of support at the U.S. Department of Agriculture's Forest Service and at the Cooperative State Research, Education, and Extension Service (CSREES)-funded wood utilization research centers from fiscal years 1995 through 2005; and (3) how the federal government transfers technologies and products from its wood utilization research and product development activities to industry.

For this review, we defined wood utilization research and product development as those activities that occur from harvesting the wood through the recycling of wood and paper products. To better understand the focus of the federal research and development efforts in wood utilization, we worked with Forest Service and CSREES program officials to develop the following five broad categories: (1) harvesting—using scientific and engineering principles to ensure cost-effective, environmentally acceptable, and safe forest operations, including planning, road building, harvesting, handling and processing, and transportation; (2) wood properties—studying the basic and applied physical, chemical, and mechanical properties of wood and wood fiber to determine the suitability of this material for various uses, from pulp to structural beams to recycled composite products; (3) manufacturing and processing—new and better manufacturing ways to extract, reduce, and convert virgin wood raw materials to useful products and the development of technology to allow the re-use of materials and products to the maximum extent possible; (4) products and testing—developing test methods and gathering and evaluating data on the differing uses of wood and wood fiber products; and (5) economics and marketing. This final category includes evaluating and tracking domestic and international supply and demand trends, and trade policies, and markets, including market opportunities; and harvesting and production costs for alternative material and energy inputs and processing options.

We performed our work at 12 federal agencies that support wood utilization research and product development activities. These include CSREES, the Forest Service, and the Natural Resources Conservation Service; the Department of Defense's (Defense) Army, Army Corps of Engineers, and the Office of Naval Research; the Department of Energy; the Department of Homeland Security's Coast Guard; the Department of Housing and Urban Development (HUD); the Department of the Interior's (Interior) Bureau of Indian Affairs; the National Science Foundation; and the Department of Transportation.

To answer the first objective—describing the types of wood utilization research and product development activities supported by federal agencies and how these efforts are coordinated—we collected information on research and product development activities at the 12 agencies for fiscal years 2004 and 2005 and worked with the Forest Service and CSREES to place these activities into one of the five categories we had developed. Because certain Forest Service research work units and CSREES-funded wood utilization research centers are specifically dedicated to wood utilization research and product development, we collected data on research activities for fiscal years 1995 through 2005 to understand how these activities changed over time. At the Forest Service, we used a data collection instrument to systematically gather data on the 27 research work units' plans for wood utilization research and product development, covering fiscal years 1995 through 2005. Because these plans span multiple years, some dated back as far as 1988. In total, we examined the 71 plans for the 16 research work units at the Forest Products Laboratory and 11 research work units that were associated with other research stations within the Forest Service—4 in the Northeast, 4 in the South, 1 in the Pacific Northwest, 1 in the Pacific Southwest, and 1 in the Rocky Mountains. From these plans, we collected information on each research work unit's mission, research problems, and selected research activities. (See app. II.) We also interviewed each research work unit's project leader on the unit's wood utilization research and product development activities.

For CSREES, we examined the 10 wood utilization research centers at 12 universities that receive congressional committee-directed grants for wood utilization research and product development. Nine of these centers are at the universities of Alaska Southeast, Minnesota-Duluth, Maine, and Tennessee; Michigan State University, Mississippi State University, North Carolina State University, Oregon State University, and West Virginia University; and the tenth center is divided among three universities—Idaho State, Montana State, and Washington State—that participate in the Inland Northwest Forest Products Research Consortium. To identify these centers' wood utilization research and product development activities, we obtained copies of the research proposals that the centers submit annually to CSREES. We used a data collection instrument to (1) systematically review the 88 proposals for fiscal years 1995 through 2005; (2) obtain information on the research objectives, approach, and description of wood utilization research and product development activities; and (3) summarize selected activities for reporting purposes. We also obtained information on the centers' research activities from CSREES' Current Research Information System (CRIS) to obtain concise, nontechnical descriptions of selected activities and to ensure that the CRIS summary reflected the information in the CSREES proposals. We interviewed knowledgeable agency officials regarding the reliability of data we used from CSREES' CRIS database and compared selected CRIS data with grant files. We used the data from CSREES for descriptive purposes only, and determined that the data were sufficiently reliable for these purposes. For reporting purposes, we primarily relied on the CRIS summary information to describe the selected research activities presented in appendix III.

To identify other CSREES wood utilization research and product development activities in fiscal years 2004 and 2005, CSREES officials queried the CRIS database using key search codes to identify the wood utilization research and product development activities being conducted under other CSREES-funded grant programs. At the time of our review, the CRIS database did not contain complete information for fiscal year 2005. We reviewed the grant projects—104—that fell within our definition of wood utilization research and product development.

To collect information on wood utilization research and product development from the remaining 10 agencies, we interviewed agency officials and reviewed and summarized available information on the research activities for fiscal years 2004 and 2005.

To obtain information on the coordination of wood utilization and product development activities among the 12 federal agencies, we interviewed agency officials to obtain their views on the use of informal and formal coordination mechanisms. For all agencies, we obtained this information through interviews with program officials and scientists. In the case of CSREES wood utilization research centers, we obtained this information through a data collection instrument sent to the program leader at each center. In addition, we obtained documents on selected formal coordinating mechanisms, such as interagency agreements. We also attended the "Agenda 2020" meeting sponsored by the

Forest Service in 2005, which is held annually to exchange information between industry and Forest Service scientists performing wood utilization research and product development activities.[1] The Forest Service uses these meetings to seek industry views on research results and future research needs. We also examined relevant laws, regulations, and agency polices related to coordination for wood utilization research and product development.

To address the second objective—describe the level of support federal agencies made available for wood utilization research and product development activities in fiscal years 2004 and 2005, and changes in the level of support at the Forest Service and CSREES wood utilization research centers from fiscal years 1995 through 2005—we collected budget authority or expenditure information from the 12 agencies for fiscal years 2004 and 2005, and from the Forest Service and CSREES' wood utilization centers for fiscal years 1995 through 2005. We reported dollars in either budget authority or expenditure data, depending on the availability of agency data. We analyzed these data in both nominal (actual) dollars and dollars adjusted for inflation (real).[2] Most agencies and programs received congressional committee-directed budget authority for wood utilization research and product development or allocated a portion of their budget authority for these activities. Those budget authority amounts are reported when available. However, the only data available for the other CSREES grants and for the National Science Foundation were expenditure data.

For information on CSREES' budget authority for the wood utilization research centers for fiscal years 1995 to 2005 for the grants awarded to the wood utilization research centers, the CSREES official explained how the funds were allocated across the 10 wood utilization research centers over the 11-year period. These data were used to show the historical trends of investment dollars for wood utilization research and product development over the past 11 years. (See app. IV.)

In addition to the budget authority for the CSREES wood utilization research centers, we obtained expenditure data for the wood utilization research and product development activities conducted under the authority of the McIntyre–Stennis Act, the Hatch Act, the National Research Initiative, the Small Business Innovation Research Grants, and other small grants, which can fund wood utilization research and product development. We obtained specific expenditure amounts for these activities for fiscal year 2004 from the CRIS database system. Fiscal year 2005 data were not available for these CSREES activities.

For the Forest Service, we obtained information on budget authority from an internal agency review of research stations and research work units. We used this information to provide an overview of the changes in budget authority for wood utilization research and product development for fiscal years 1995 through 2005. See appendix IV for the budget authority for each research work unit over this period. In addition, we interviewed Forest Service budget officials in headquarters, the Forest Products Laboratory, and the State and Private Forestry Program on budget and other funding issues, such as the allocation of funds and setting of research funding priorities. We concluded that the data provided in the agency survey were sufficiently reliable for the purposes of our review.

We also reviewed and summarized information from Forest Service documents on the number of scientists and research support staff at the Forest Service—the only agency that has full-time federal employees who directly conduct wood utilization research and product development activities. We reported the number of full-time equivalent (FTE) staff at each of the 27 research work units that conducted research on wood utilization and product development for fiscal years 1995 through 2005. (See app. IV.)

To collect funding information from the remaining agencies, we asked budget and program officials for budget authority or expenditure information for fiscal years 2004 and 2005 for wood utilization research and product development. Specifically, the National Science Foundation provided us with expenditure information from its Project Reports Summary and Search and Awards databases because that is the only way it could identify the amounts devoted to wood utilization research and product development. We interviewed knowledgeable agency officials regarding the reliability of these data. We used the data for descriptive purposes only, and determined that the data were sufficiently reliable for these purposes.

The funding for Defense's Army, Corps of Engineers, and Office of Naval Research; and the Department of Homeland Security's Coast Guard were congressional committee-directed funds or budget authority. However, for the Office of Naval Research and the Coast Guard, we reported expenditures because those amounts were applicable to our time period—fiscal years 2004 and 2005.

To respond to objective three—how the federal government transfers technologies and products from its wood utilization research and product development activities to industry—we obtained and reviewed relevant legislation and policies and procedures on federal technology transfer activities. At the Forest Service, we interviewed and obtained examples of successful technology transfer from project leaders at the 27 research work units that are responsible for wood utilization research and product development; a patent attorney; technology transfer program managers at the Technology Marketing Unit located at the Forest Products Laboratory; and technology transfer specialists in the State and Private Forestry Program. At CSREES, we had discussions with program research officials and extension specialists. In addition, we sent a short data collection instrument to the 10 wood utilization research centers to obtain information on how they transfer the results of their research to industry, as well as to obtain examples of successful transfer efforts. We did not assess the success of these agencies' reported efforts, nor did we try to quantify the results of these efforts.

We also conducted site visits at a limited number of federal, university, and industrial facilities—the Forest Products Laboratory; Forest Service facilities in Virginia, West Virginia, and Oregon; the wood utilization research center at Oregon State University; the Western Wood Producers Association; the APA Engineered Wood Association; and a Weyerhaeuser Company research laboratory in Washington State. We also visited a sawmill, a manufacturer of wooden steps and stair posts, a manufacturer of engineered products, and a cabinet maker, and attended the 2005 Northeast Utilization and Marketing Council's conference. We performed our work between February 2005 and May 2006, in accordance with generally accepted government auditing standards.

REFERENCES

[1] The official title of this meeting is "Forest Industry and Forest Service Research Liaison Meeting." The meeting was held at the Forest Products Laboratory in Madison, Wisconsin, May 10-11, 2005.

[2] We adjusted nominal dollars using the Department of Commerce's Fiscal Year Chain-Weighted Price Index for the Gross Domestic Product with 2004 as the base year.

In: Wood Utilization
Editor: J. E. Carter, pp. 53-108
ISBN: 1-60021-459-2

APPENDIX II: FOREST SERVICE'S CONDUCTED AND PLANNED WOOD UTILIZATION RESEARCH AND PRODUCT DEVELOPMENT

This appendix presents examples of work conducted and planned for the Forest Service's research work units at the Forest Products Laboratory (table 14), and in work units associated with five research stations: Northeastern, Pacific Northwest, Pacific Southwest, Rocky Mountain, and Southern (table 15).

Table 14. Conducted and Planned Activities for the Forest Service's Research Work Units at the Forest Products Laboratory

Forest Products Laboratory's research work unit and plan components			
	Research work unit title, and period and activities covered		
4502			
	Biodeterioration of Wood, 1991-97	Biodeterioration of Wood, 1997-2002	Biodeterioration of Wood, 2002-07
Area of research applicability	National	National	National
Mission	To increase wood efficiency and serviceability through basic and applied research on the nature and control of biodeterioration of wood in use.	To increase wood efficiency, protection, and serviceability through basic and applied research on the nature and control of biodeterioration within the context of changing environmental needs.	To increase wood efficiency of use, protection, and serviceability through basic and applied research on the nature and control of biodeterioration within the context of changing environmental needs.
Research problem and activities	*Problem*: Lack of understanding of the biosynthetic and degradative pathways that can be targeted to achieve improved methods for controlling decay. *Activities*: Study the induction and repression of degradative enzymes in order to learn what parameters prevent their formation. *Problem*: Lack of understanding in detecting incipient decay and in-situ treatments to protect wood.	*Problem*: Need to identify new biochemical processes of wood decay fungi that can be inhibited to achieve improved methods for controlling decay. *Activities*: Determine efficacy against decay fungi of inhibitors to targets identified from mechanistic studies. *Problem*: Need to identify in-situ modes of antagonism of nondecay microorganisms required to enhance their establishment in wood to prevent fungal attack in order to develop new and effective biocontrol agents.	*Problem*: Need to identify new specific biochemical processes of wood decay fungi and mold that can be inhibited to achieve improved methods for controlling decay. *Activities*: The recent sequencing of the entire genome of an important decay fungus presents many opportunities for research on the mechanism of decay, including the possibility of rapidly correlating specific enzymes of the fungus with the corresponding genes.

Table 14. (Continued)

Forest Products Laboratory's research work unit and plan components			
	Research work unit title, and period and activities covered		
	Activities: Investigate the presence of decay in a particular wood structure to determine conditions promoting such decay and, where necessary, the microorganisms responsible for the decay. *Problem*: Lack of understanding in how to enhance nondecay microorganisms to prevent fungal attack of wood. *Activities*: Identify potential antagonists by laboratory screening and select the most promising control agents by applying them to wood blocks for exposure to important wood-attacking fungi.	*Activities*: Identify microorganisms with improved biocontrol qualities. *Problem*: Need to determine the conditions required for bioremoval of metals from chromated copper arsenate (CCA)-treated wood waste in order to dispose of CCA-treated wood without damaging soils and watersheds. *Activities*: Develop a fuller understanding of the microbial degradation of CCA-treated wood wastes and the conditions required for remediation, recycling, or composting.	*Problem*: Need to determine the conditions required for bioremoval and recycling of preservatives from preservative-treated wood waste in order to safely dispose of preventative-treated wood without damaging soils and watersheds. *Activities*: Develop a fuller understanding of the microbial remediation and degradation of preservative-treated wood wastes and the conditions required for remediation, recycling, or degradation (composting). *Problem*: Need to develop rapid detection methods and the conditions required for mold growth in order to prevent mold growth in housing. *Activities*: Develop improved moisture control parameters to prevent the establishment of mold and spread of mold spores.

Table 14. (Continued)

4701			
	Center for Wood Anatomy Research, 1992-98	Center for Wood Anatomy Research, 1998-2003	Center for Wood Anatomy Research, 2003-08
Area of research applicability	National	National	National
Forest Products Laboratory's research work unit and plan components			
	Research work unit title, and period and activities covered		
Mission	To accumulate and make known information on the anatomical and other characteristics of woods of the world that may affect their utilization potential, and to develop new and improved techniques for their identification.	To accumulate and disseminate information on the anatomical, biochemical, and physical characteristics of wood species that may affect their utilization and decay, and to develop new and improved techniques for wood identification.	To develop, accumulate, and disseminate information on the anatomical, biochemical, and physical characteristics of wood species that may affect their utilization and wood quality, and to develop new and improved techniques for wood identification.
Research problem and activities	*Problem*: Need to develop a complete body of fundamental knowledge on the macro- and microscopic structures and other characteristics for wood to better understand, utilize, and identify lesser-known and lesser-used woods, especially tropical woods. *Activities*: Data-gathering and analysis that is directed toward the development of descriptions and keys for commercially important and emerging tropical groups.	*Problem*: Need to better understand the macro- and microscopic anatomy of wood and develop new techniques and methodologies for wood identification. *Activities*: Data-gathering and analysis that is directed toward the development of descriptions and keys for commercially important and emerging tropical groups.	*Problem*: Need to better understand the macro- and microscopic anatomy of wood and to develop new techniques and methodologies for wood identification. *Activities*: Data-gathering and analysis that is directed toward the development of descriptions and keys for commercially important and emerging tropical groups.

Table 14. (Continued)

Forest Products Laboratory's research work unit and plan components	Research work unit title, and period and activities covered		
	Problem: Need to investigate new and improved nonanatomical techniques and methodologies for wood identification for more accurate and reliable identifications. *Activities*: New and novel approaches to the separation of similar woods must be developed. One of the first techniques to be studied will be unique chemical tests that might aid in identifying woods at the work site. *Problem*: Need to accumulate common and scientific names, origin, physical and strength properties, uses, and other information on tropical species in databases to assist in organizing and disseminating information to customers. *Activities*: Gather information on properties, uses, and other data on native and tropical species, convert it to a standard format, and develop database structures that can readily retrieve selected information in an orderly and efficient manner.	*Problem:* More information is needed to understand the interrelationships between specific anatomical and chemical features and the mechanisms of decay and natural durability. *Activities*: Identify and test selected wood species for the ability to resist decay and discoloration by brown-rot, white-rot, and soft-rot fungi. Emphasis will be given to laboratory testing of selected temperate and tropical species. *Problem*: Need to understand the relationships between wood anatomical characteristics and wood quality. *Activities*: In the past, several characteristics (mostly anatomical) have been investigated as predictors of wood quality. These include growth rate (ring width), density/specific gravity, percentage of latewood, tracheid length, cell diameter, cell wall thickness, and cellulose microfibril angle. Cellulose microfibril angle contributes to many wood properties, such as modulus of elasticity, creep, shrinkage, and maximum crushing strength.	*Problem*: Need to understand the relationships between wood anatomical characteristics and wood quality. *Activities*: Develop better methods for delineating juvenile zones, and conduct closer comparisons of juvenile and mature wood anatomy and properties to better assess the impact of juvenile wood on overall wood quality.

Table 14. (Continued)

Forest Products Laboratory's research work unit and plan components	Research work unit title, and period and activities covered		
4703			
	Wood Adhesives Science and Technology, 1994-99	Wood Adhesives Science and Technology,1999-2004	Wood Adhesives Science and Technology, 2004-09
Area of research applicability	National	National	National
Mission	Improve the utilization of wood through a combination of basic and applied research that will ensure adequate future supplies of durable, environmentally acceptable adhesives, and improve the applicability, efficiency, and durability of adhesives for bonding wood-to-wood and wood-to-nonwood materials.	Improve the utilization of wood through a combination of basic and applied research that seeks to ensure more efficient fabrication and performance of bonded-wood products.	Improve the utilization of wood through a combination of basic and applied research that ensures more efficient fabrication and performance of bonded-wood products.
Research problem and activities	*Problem:* Need to develop durable wood adhesives from renewable resources. *Activities*: Utilize the chemical constituents that occur in wood and other biomass (or compounds derived from these constituents) for obtaining adhesive systems suitable for bonding wood products.	*Problem*: Need to improve the properties and performance of wood adhesive systems. *Activities*: Demonstrate the use of computational chemistry methods to develop a better fundamental understanding of the chemistry involved in the synthesis and cure of existing adhesive systems.	*Problem*: Need to improve the durability, safety, and structural performance of adhesively bonded wood assemblies. *Activities*: Develop a methodology to understand where and why failure is taking place when bondline failure occurs.

Table 14. (Continued)

Forest Products Laboratory's research work unit and plan components	Research work unit title, and peri d and activities covered		
	Problem: Need to eliminate chemical and physical barriers to bonding wood- based materials. *Activities*: Identify and develop means to overcome chemical and physical barriers to bonding and wood-based materials that have been chemically or physically altered through treating or processing, among other things. *Problem*: Need to improve the environmental acceptability of wood adhesives through the modification of conventional adhesives and the development of new adhesives. *Activities*: Quantify the magnitude of the environmental impact of composite wood panel production and use and develop methods to decrease the environmental impact of the composite products.	*Problem*: Need to improve the adhesion of adhesives to wood and nonwood substrates. *Activities*: Understand the factors that lead to poor wood-surface quality and their influence on wood bonding so that the most efficient, cost-effective bonding solutions can be determined. *Problem*: Need to improve the durability, safety, and structural performance of adhesively bonded wood assemblies. *Activities*: Determine the mechanical behavior of adhesives within bonded materials and develop new information on the mechanical properties of adhesively bonded joints. *Problem*: Need to understand the environmental acceptability of wood adhesives and composite wood panels.	*Problem*: Need to better understand the adhesion of adhesives to wood and nonwood substrates. *Activities*: Understand the factors that lead to poor wood-surface quality and their influence on wood bonding so that the most efficient, cost-effective bonding solutions can be determined. *Problem*: Need to improve the properties and performance of wood adhesive systems. *Activities*: Determine the mechanical properties of existing adhesive systems. *Problem*: Need to utilize more environmentally acceptable wood adhesives in wood composite panels. *Activities*: Evaluate the VOC emissions from composite wood products.

Table 14. (Continued)

Forest Products Laboratory's research work unit and plan components	Research work unit title, and period and activities covered		
	Problem: Need to determine the behavior of adhesives and the performance of adhesive-bonded assembly joints in wood structures used for construction. *Activities*: Develop knowledge that will support and encourage the use of adhesives in building construction, opening the way for improvements in the use of wood and the performance of wood structures.	*Activities*: Develop information on volatile organic compounds (VOC) emitted from wood products. Successful completion of this component will provide the information needed by other researchers, manufacturers, and regulatory agencies to determine the impacts that wood products have on indoor air quality, and to develop strategies to control or prevent exposure to VOCs.	
4706			
	Performance Designed Composites, 1993-97	Performance Designed Composites, 1997-2002	Performance Designed Composites, 2002-07
Area of research applicability	National	National	National
Mission	To develop the capabilities in processing technology to improve the value and yield of existing (and new) lignocellulosic-based composites.	To conserve wood and other renewable fiber resources, this research unit determines the relationship between alternative resource options, processing technologies, and composite performance levels to improve the value and yield of existing and new lignocellulosic-based composites.	To conserve wood, alone or in combination with other renewable fiber resources, fundamental relationships between base materials and product performance are defined and then processes are derived to engineer reliable, high-performance composites from wood- and wood-lignocellulosics, including new hybrid composites melding wood and alternative materials.

Table 14. (Continued)

Forest Products Laboratory's research work unit and plan components	Research work unit title, and period and activities covered		
Research problem and activities	*Problem*: There is a lack of knowledge that relates the resource and processing variables to composite product performance. *Activities*: Study alternative material options and fabrication techniques that result in flat or shaped products that can either span distances of 4 feet or more or that possess other unique properties. *Problem*: There is a lack of information available on how to economically produce and characterize the performance of inorganic bonded structural composites in adverse environments. *Activities*: Develop a broad database of processing and performance information, which will make possible the introduction of economically produced and performance-characterized inorganic bonded composites into the U.S. marketplace.	*Problem*: Knowledge is needed to characterize, assess, and prepare alternative raw materials for processing into value-added composite products. *Activities*: Develop correlations between raw material influences, composite design, the physical properties of the constituent biomass components, adhesive bonding mechanisms, and the manipulative variables of product fabrication. *Problem*: Need to develop new composite processing technologies and to refine existing technologies to ensure that the composite raw materials of the future are optimally assembled to achieve maximum performance at minimum cost. *Activities*: Develop a database of information to define and characterize various raw materials and processing options that can affect the performance of alternative composites.	*Problem*: It is necessary to understand the fundamental relationships between wood-, natural-fiber- and alternative materials and use this knowledge to optimize composite processing and performance. *Activities*: Develop correlations between raw material influences, composite design, the physical properties of the constituent biomass components, adhesive bonding mechanisms, and the manipulative variables of product fabrication. *Problem*: Need to improve the performance, durability and value of existing composites and define the next generation of hybrid composites made from natural fiber(s) and alternative materials.

Table 14. (Continued)

Forest Products Laboratory's research work unit and plan components	Research work unit title, and period and activities covered		
	Problem: There is a lack of information to allow the development of processing methodology for the production of high-performance composites from virgin and recycled lignocellulosic, plastic, and other nonwood materials. *Activities*: Develop technology to convert recycled biofibers and nonwood materials into durable, long-service-life products that are recyclable and otherwise environmentally friendly, and will effectively remove raw materials from the waste stream.	*Problem*: Information is needed to characterize, predict, and correlate composite performance based on raw material, processing, and structure considerations. *Activities*: Develop technology to convert wood and woody biomass, recycled, and nonwood materials into durable, cost-effective, high-performing and long-service-life products that are recyclable and otherwise environmentally friendly.	*Activities*: Build upon the fundamental knowledge gained in the above activities to develop a database of information to define and characterize various raw materials and processing options that can improve the performance of traditional composites (or define new composites). *Problem*: Need to develop tools to address resource sustainability, enhance recyclability, and minimize the environmental impacts of composite processing. *Activities*: Optimize how wood and woody biomass, recycled and nonwood materials are converted into durable, cost-effective, high- performing, and long-service-life products that are recyclable and otherwise environmentally friendly.
4707	Wood Surface Chemistry, 1994-99	Wood Surface Chemistry, 1999-2004	Wood Surface Chemistry, 2004-09
Area of research applicability	National	National	National
	To determine the basic mechanisms of wood surface deterioration outdoors—and innovative products and processes—for modifying wood surfaces to enhance finishability and gluability and to improve properties of wood and nonwood composites.	To determine the basic mechanisms of deterioration of wood and wood-based composites used outdoors and develop innovative technologies for modifying wood surfaces to enhance durability.	To improve the durability of wood and wood-based composites.

Table 14. (Continued)

Forest Products Laboratory's research work unit and plan components	Research work unit title, and period and activities covered		
Research problem and activities	*Problem*: Rapid changes in wood and wood-based materials, products used on these materials, and the environment, have led to premature failure of many wood products used outdoors. *Activities*: Address chemical changes on the surface of wood and between the paint and wood interface and evaluate paint adhesion so that the chemical changes can be linked to measurable paint performance parameters. *Problem*: Inadequate wetting and penetration of wood finishes, adhesives, and other treatments cause decreased service life of many wood products. *Activities*: Research will concentrate on the basic surface properties of wood and wood-based materials as they relate to interaction with liquids. *Problem*: Chemical incompatibilities at the wood/nonwood interface do not permit the development of high-performance composites.	*Problem*: Premature weathering and decay of wood products used outdoors causes an unnecessary drain on our forest resource. *Activities*: Address chemical changes on the surface of wood and at the paint/wood interface and evaluate paint adhesion so that the chemical changes can be linked to measurable paint performance parameters. *Problem*: Water-based water-repellent preservatives are not protecting products as well as traditional solvent-based formulations. *Activities*: Research will concentrate on the basic surface properties of wood and wood-based materials as they relate to interaction with liquids. *Problem*: Incompatibility of surface interactions between wood and other materials impedes the development of advanced wood-based composites.	*Problem*: Wood and wood-based materials used in residential construction are not performing as expected because of inadequate understanding of the mechanisms by which they are failing and inadequate methods for predicting their service life. *Activities*: Address chemical changes on the surface of wood and at the interface with other materials and also evaluate adhesion so that the chemical changes can be linked to measurable performance parameters. *Problem*: Lack of understanding of the surface interactions between water and wood or other lignocellulosic materials limits their use in many traditional and new applications. *Activities*: Research will concentrate on the basic surface properties of wood and wood-based materials as they relate to interaction with liquids and solutions.

Table 14. (Continued)

Forest Products Laboratory's research work unit and plan components	Research work unit title, and period and activities covered		
	Activities: Attempt to elucidate the complex chemical structure of wood/nonwood interphases and to modify the chemistry of the wood surface to achieve better bonding between wood and nonwood materials.	*Activities*: Attempt to elucidate the complex chemical structure of wood/nonwood interphases and to modify the chemistry of the wood fiber surface to achieve better bonding between wood and nonwood materials. The major emphasis of this research is the measurement of wood surface properties critical to good bonding.	*Problem*: Inadequate understanding of the surface chemistry of wood and the mechanism by which other materials bond to wood impedes the development of advanced wood-based composites. *Activities*: Elucidate the complex chemical structure of wood/nonwood interphases and develop processes to modify the chemistry of the wood fiber surface to achieve better bonding between wood and nonwood materials.
4709	**Chemistry and Pulping, 1991-97**	**Chemistry and Pulping, 1997-2002**	**Chemistry and Pulping, 2002-07**
Area of research applicability	National	National	National
Mission	To combine state-of-the-art knowledge in wood chemistry and chemical engineering science in developing environmentally benign processes for the production and utilization of wood pulp fibers and the chemical byproducts of pulping processes. The new processes will seek to conserve forest and water resources, avoid adverse effects on air quality and the global ecosystem, and reduce capital investment in order to enhance the competitiveness of U.S. pulp and paper products in world markets.	To develop environmentally benign and resource-conserving processes for the production and utilization of wood pulp fibers, and of the chemical byproducts of wood and pulp processing, and to improve our understanding of the molecular and physical characteristics of wood and wood pulp in order to achieve a sustainable basis for the production of U.S. pulp and paper products to meet the needs of the American people.	To develop more efficient, environmentally benign, and resource-conserving processes for the conversion of wood to fibers and chemicals, and to improve our understanding of the chemical, molecular, and physical characteristics of wood and fibers to provide a basis for sustainable conversion of wood into value-added products.

Table 14. (Continued)

Forest Products Laboratory's research work unit and plan components	Research work unit title, and period and activities covered		
Research problem and activities	*Problem*: Need to develop new bleaching technologies. *Activities*: Identify the major barriers to developing alternative bleaching systems (to chlorine-based systems), and target fundamental research programs to overcome these barriers. *Problem*: Brightness reversion in high-yield pulps. *Activities*: Develop information to provide a basis for addressing the problem of brightness reversion and address the gap in knowledge about the phenomena responsible for reversion. *Problem*: New pulping concepts. *Activities:* Seek systems which rely on catalysts to break down the lignin, rather than requiring severe thermal and chemical environments implicit in high-temperature alkaline pulping, and developing systems that are based on simulating the action of biological systems that break down lignin.	*Problem*: New technologies are needed to deliqnify wood for the production of pulp in ways that will eliminate contamination of our air and waterways. *Activities*: Continue to define the fundamental science and engineering that must be understood in order to fully develop the potential of polyoromeralate delignification for effluent-free oxygen bleaching. *Problem*: Need for improved papermaking properties of high-yield mechanical and chemi-mechanical pulps, which makes more efficient use of our wood resource; increase public acceptance of, and build markets for papers, produced from these pulps. *Activities*: Develop information to provide a basis for addressing the problem of brightness reversion and to develop brightness-stabilizing procedures.	*Problem*: New technologies are needed to fractionate wood for the production of pulp in ways that will eliminate contamination to our air and waterways. *Activities*: Further develop the potential of polyosometalate delignification to provide solutions to the environmental problems currently associated with delignification processes. *Problem*: Need to improve papermaking of high-yield mechanical and chemi-mechanical pulps, which make more efficient use of our wood resources, to increase public acceptance of and build markets for papers produced from these pulps. *Activities*: Develop information to provide a basis for addressing the problem of brightness reversion and to develop brightness-stabilizing approaches. *Problem*: Need for new and innovative methods to convert wood and other lignocellulosics into fibers and chemicals.

Table 14. (Continued)

Forest Products Laboratory's research work unit and plan components	Research work unit title, and period and activities covered		
	Problem: Recycling. *Activities*: Research programs addressing the problem of surface hardening will need to focus on new approaches to reactivating the surface to enhance interfiber bonding. *Problem*: High-value chemicals from wood. *Activities*: Continue some efforts directed at the development of analytical procedures currently underway and develop a new effort with a focus on the problem of color in certain products. *Problem*: Characterization of wood components. *Activities*: Studies of the molecular architecture of cell walls in wood, and the manner in which it is transformed by the various biological and industrial processes which break down the native structures using novel methods for characterizing the solid state and states of molecular aggregation.	*Problem*: Need for increased understanding of the biogenesis and molecular architecture of wood cell walls, their response to environmental stresses, and their transformation in the course of industrial processing to improve forest health, utilize mixed species of uneven acres, and increase efficiency of conversion. *Activities*: Studies of the molecular architecture of cell walls in wood and the manner in which it is transformed by the various biological and industrial processes that break down the native structures.	*Activities*: Develop technologies for biorefining wood into ethanol and other chemicals, fiber, and structural materials in much higher yields than are currently possible. *Problem*: Need for increased understanding of the biogenesis and molecular architecture of wood cell walls, their response to environmental stresses, and their transformation in the course of industrial processing to improve forest health, utilize mixed species of uneven age, and increase efficiency of conversion. *Activities*: Studies of the molecular architecture of cell walls in wood and its relationship to the process of biogenesis.

Table 14. (Continued)

4710	Fiber Processes and Products, 1990-97	Fiber Processing and Paper Performance, 1997-2002	Fiber Processing and Paper Performance, 2002-07
Areas of research applicability	National	National	National
Mission	To improve the efficiency with which pulps derived from the nation's wood and recycled fiber resources are converted to fiber-based products.	To conserve fiber resources by developing the knowledge and technology needed to better utilize a fiber supply from a wide range of biomass resources, increase the use of recycled fiber, improve paper performance, and address environmental concerns.	To conserve forest resources through paper performance research aimed at increasing the use of small-diameter and underutilized tree species, recycled fiber, and a wide range of biomass resources addressing environmental and energy concerns.
Forest Products Laboratory's research work unit and plan components	Research work unit title, and period and activities covered		
Research problem and activities	*Problem*: Need to improve papermaking processing to reduce fiber needs and increase the use of underutilized fiber sources. *Activities*: Develop ways to permit basis weight to be reduced without decreasing strength and opacity. An approach to this problem would be to apply press-drying principles. *Problem*: Need to develop the technologies to increase the use potential of high-yield mechanical pulps from a wide variety of hardwood and softwood species.	*Problem*: New technologies are needed to better utilize fiber from a wide range of biomass resources for pulp and papermaking to improve forest health and conditions to utilize wood now going to waste streams. *Activities*: Mechanical pulps are currently made from a small number of select softwood species, and only one or two low-density hardwoods. There are some indications that juvenile wood, which predominates in thinnings and small-diameter trees, might be advantageous for mechanical pulp production.	*Problem:* Basic knowledge and technology is needed to increase the use of small-diameter and underutilized tree species for pulp and paper products to improve forest health and reduce fire risk. *Activities*: Substituting mechanical pulp for chemical pulp is one way to extend the current wood supply. There are some indications that juvenile wood, which predominates in thinnings and small-diameter trees, might be advantageous for mechanical pulp production.

Table 14. (Continued)

Forest Products Laboratory's research work unit and plan components	Research work unit title, and period and activities covered		
	Activities: Research will be conducted to minimize reduction in optical properties whenever strength properties are increased. *Problem*: Need to develop the technologies to increase the use of recycled wastepaper. *Activities*: Use a materials-science approach to develop a better understanding of the role of solid-liquid interfacial phenomena in separation of synthetic adhesive contaminants from wastepaper pulp slurries. *Problem*: Need to improve product performance of paper and the efficient use of fiber resources through chemical treatments and incorporation on nonwood components.	*Problem*: New technologies are needed to overcome undesirable environmental impacts in converting wood to paper and paperboard. *Activities*: Investigate the use of fungal pretreatment for kraft pulping. Fungal pretreatment of the wood chips enhances the strength properties of the paper while reducing the toxicity of the waste stream for mechanical and sulfite pulping, as well as reducing the consumption of mechanical pulping energy. *Problem*: New recycling technologies are needed to overcome barriers to increased use of wastepaper. *Activities*: The technology for using enzymes to remove ink from toners needs to be transferred to industrial practice. The recycling of paper into pulp suitable for papermaking requires numerous steps of several unit operations to produce an acceptable product..	*Problem*: New high-yield pulping technologies are needed to reduce energy consumption, improve paper quality, and overcome undesirable environmental impacts in converting wood to paper and paperboard. *Activities*: Investigate enzyme-assisted grafting of carboxylic acid groups on the surface of lignin-containing pulp fibers as a post treatment for biotreated thermal-mechanical pulps. *Problem:* New technologies are needed to overcome barriers to increased recycling of recovered papers. *Activities*: New approaches are needed to better use fibers from recovered paper and paper mill residues. The major contaminant categories are inks, adhesives, plastics, and inorganic materials.

Table 14. (Continued)

Forest Products Laboratory's research work unit and plan components	Research work unit title, and period and activities covered		
	Activities: Part of the solution to the problem of insufficient opacity of low-basis weight papers involves the use of nonfiber components as fillers and opacifiers.	*Problem*: Greater knowledge is needed to overcome limits in our fundamental understanding of the relationship between fiber properties and paper performance to optimize fiber use and extend forest resources *Activities*: An improved understanding of how the performance of corrugated containers relates to paper properties can provide the rationale to differentiate among alternative fiber sources.	*Problem*: Need for a fundamental understanding of the relationship between fiber properties and paper performance to optimize fiber use and extend forest resources. *Activities*: An improved understanding of how the performance of corrugated containers relates to paper properties can provide the rationale to differentiate among alternative fiber sources.
4712	Institute for Microbial and Biochemical Technology, 1991-97	Institute for Microbial and Biochemical Technology, 1992-2002	Institute for Microbial and Biochemical Technology, 2002-07
Area of research applicability	National	National	National
Mission	To explore, through basic and applied research, the potentials of biotechnology in wood conversions.	To explore, through basic and applied research, the potentials of biotechnology in wood conversions, and to better understand the roles of such conversions in enhancing our environment.	Develop biotechnology for wood and fiber conversion through fundamental and applied research that contributes to efficient utilization and improved health of our forests.
Research problem and activities	*Problem*: Need to understand the aromatic-mineralizing system of lignin-degrading fungi to permit its use in many applications, as well as its appreciation as a key component of the earth's carbon cycle.	*Problem*: Need to better understand oxidative systems in wood decay fungi to permit their use in many applications, including improvement of forest health.	*Problem*: Need for technologies for efficient and economical bioconversion of forest thinnings, unmerchantable timber and mixed species to produce a full array of wood-based materials ranging from fiber to constitutive polymers and chemicals.

Table 14. (Continued)

Forest Products Laboratory's research work unit and plan components	Research work unit title, and period and activities covered		
	Activities: Basic investigations into the mechanism of lignin degradation by selected fungi. *Problem*: Need for a greater understanding of the biochemistry, physiology, and engineering of xenobiotic degradation by lignin-degrading fungi to allow use of these organisms in bioremediation technologies. *Activities*: Gain an understanding of fungal physiology, fungal ecology, and microbial engineering. *Problem*: Need to identify the best enzymes and necessary conditions for their action to permit their use in improving properties of virgin and recycled fibers. *Activities*: Basic and applied studies on the use of microbial enzymes to treat virgin and recycled fibers. *Problem*: Need to evaluate completely the promising concept of biopulping to determine its technical feasibility.	*Activities*: Basic investigations into the mechanisms of lignin and cellulose degradation by selected white rot and brown rot fungi. *Problem*: Research is needed to provide basic information and operating conditions for enzymatic processing of wood fiber. *Activities*: Basic and applied studies on the discovery and use of microbial and low-grade enzymes to treat virgin and recycled fibers, wood residues from timber harvest, or excess growth.	*Activities*: Increase the efficiency of bioconversion to ethanol by engineering the metabolic pathways for sugar utilization and fermentation. *Problem*: Need to identify and understand the oxidative systems of wood decay fungi to increase the efficacy of these naturally occurring organisms in bioconversion technologies, devising new environmentally sound ways to protect wood in use, and accelerating decomposition of forest litter.

Table 14. (Continued)

Forest Products Laboratory's research work unit and plan components	Research work unit title, and period and activities covered		
	Activities: Biopulping research is divided into research teams, each charged with a specific part of the overall effort: fungal, pulp and paper, engineering and scale-up, enzyme, and molecular genetics teams. *Problem*: Need to develop fundamental knowledge concerning the molecular genetics of ligninolytic fungi for biotechnical exploitation. *Activities*: Elucidate the structure, organization, and regulation of fungal genes involved in lignocellulose degradation. *Problem*: Lack of fundamental knowledge of the enzymes and conditions necessary for producing fermentable hydrolystes; and need to improve microbial strains before fermentation of the principal hemicellulosic sugars is economical. *Activities*: Examine the effects of microbial cellulases and hemicellulases on pretreated wood and lignocellulosic residues with the objective of maximizing sugar yields and concentrations.	*Problem*: Need fundamental knowledge of the molecular genetics of lignocellulose-degrading fungi for their optimal biotechnological use. *Activities*: Identify and characterize key genes involved in the degradation of lignin and related aromatic hydrocarbons, cellulose, and hemicellulose. *Problem*: Need to improve processing and fermentation technology to convert low-grade lignocellulosic materials into fuels and chemicals. *Activities*: Examine the effects of microbial cellulases and hemicellulases on pretreated wood and lignocellulosic residues with the objective of maximizing sugar yields and concentrations.	*Activities*: Basic investigations into the mechanisms of lignin and cellulose degradation by selected white rot and brown rot fungi. *Problem*: Need to develop basic information and operating conditions for efficient and effective enzymatic and microbial processing of wood and wood fiber. *Activities*: Basic and applied studies on the discovery and use of microorganisms and enzymes to treat virgin fibers, recycled fibers, and wood residues. *Problem*: Need to understand the functional genomics of lignocellulose-degrading fungi to optimize these organisms for bioprocessing and bioconversion of wood. *Activities*: Identify and characterize key genes involved in the degradation of lignin and related aromatic hydrocarbons, cellulose, and hemicellulose.

Table 14. (Continued)

4714	Engineering Properties of Wood, 1991-98	Engineering Properties of Wood, 1998- 2004	Engineering Properties and Structures, 2004-09
Forest Products Laboratory's research work unit and plan components	Research work unit title, and period and activities covered		
Area of research applicability	National	National	National
Mission	To establish methods and data for the improved characterization of the physical and mechanical properties of wood that are important in engineering design.	As part of the Forest Products Laboratory's mission to conserve wood and forest resources, the mission of the research work unit is to establish methods and gather data for the improved characterization of the mechanical and physical properties of solid sawn and composite structural products that are important in engineering design.	To contribute to the conservation and productivity of the nation's forest resources by establishing methods and gathering data for the improved characterization of the mechanical and physical properties of solid sawn and composite structural products that are important in engineering design and to foster their efficient utilization in wood building systems.
Research problem and activities	*Problem*: Physical and mechanical properties of clear wood. *Activities*: Investigate wood quality effects on properties and broadening of fundamental property data required as input for analytical modeling of material or structural performance. *Problem*: Properties of commercial lumber.	*Problem*: Properties and grading of solid-sawn lumber. *Activities*: Extend the nation's forest resource by developing and applying better grading procedures to structural lumber. *Problem*: Properties of structural composite products. *Activities*: Extend the nation's forest resources by developing analytical models and gathering data for predicting the end-use performance of composite structural products.	*Problem:* Properties and grading of solid-sawn products. *Activities*: Promote public safety and support management of the nation's forest resources by developing and applying better grading and property assignment procedures for solid-sawn structural wood products. *Problem*: Properties of structural composite products.

Table 14. (Continued)

Forest Products Laboratory's research work unit and plan components	Research work unit title, and period and activities covered		
	Activities: Focus over the next 2 years on getting the last of the descriptive information published. *Problem*: Principles of grading and property assignment for solid sawn and composite lumber. *Activities*: Develop optimum techniques for sorting lumber, predicting its mechanical properties, and assuring the reliability of these products in practice. *Problem*: Processing, treating, and environmental influences on design properties of lumber. *Activities*: Develop scientific and technical understanding of the effects of processing, treating, and environmental factors of the mechanical properties of lumber.	*Problem*: Serviceability of solid and composite products. *Activities*: Extend the nation's forest resource by developing better methods for characterizing the end-use performance of structural lumber products as a function of environmental and industrial treatment processes. *Problem*: Fundamental research. *Activities*: Develop data and procedures for characterizing the fundamental physical and mechanical properties of wood as a basis for use in other problems, and for use by researchers outside the work unit.	*Activities*: Promote public safety and support management of the nation's forest resources by developing better methods for predicting the field performance of engineered wood composites as a function of their constitutive components. *Problem*: Serviceability of solid sawn and composite wood products. *Activities*: Improve public safety and support management of the nation's forest resources by developing better methods for characterizing the end-use performance of structural products as a function of environmental change, time-dependent behavior, and industrial treating processes. *Problem*: Wood drying and heat sterilization. *Activities*: Develop economical wood drying strategies aimed at promoting value-added uses for small-diameter softwood and low-value, underutilized hardwood timber, with both the established lumber industry and small, rural, community-based businesses as the targeted users of the technology.

Table 14. (Continued)

Forest Products Laboratory's research work unit and plan components	Research work unit title, and period and activities covered		
			Problem: Properties of nontraditional wood products used in structural applications. *Activities*: Characterize the properties of wood products for safe and efficient design of structures and efficient use of the wood resource. *Problem*: Structural performance of wood-building systems. *Activities*: Better understand the structural performance of wood-building systems to better understand the use of traditional materials in residential, commercial, and industrial buildings and provide a better basis for using wood composites, reused, and recycled materials. *Problem*: Performance of wood transportation systems.

Table 14. (Continued)

Forest Products Laboratory's research work unit and plan components	Research work unit title, and period and activities covered		
			Activities: Improve existing wood transportation structures, develop new systems that conserve and improve wood use and improve the adequacy and condition of the U.S. transportation infrastructure. *Problem*: Fundamental research. *Activities*: Develop data and procedures for characterizing the fundamental physical and mechanical properties of wood as required for use in the other problem areas, and for use by researchers outside of the work unit.
4716	Engineered Wood Products and Structures, 1992-99	Engineered Wood Products and Structures, 1999-2005	Building Moisture and Durability, 2005-10
Area of research applicability	National	National	National
Forest Products Laboratory's research work unit and plan components	Research work unit title, and period and activities covered		
Mission	To extend the wood resource through engineering technology for effective design and use of wood and wood-based materials in structures.	To conserve the wood resource through engineering technology and contribute to effective design and use of wood and wood- based materials in structures.	To extend the service life of wood products in buildings through improved building design and operation.

Table 14. (Continued)

Forest Products Laboratory's research work unit and plan components	Research work unit title, and period and activities covered		
Research problem and activities	*Problem*: Engineered wood products. *Activities*: Gain the knowledge necessary to use reliability-based design, which primarily involves structural properties; however, the effect of moisture content changes on dimensional stability is also a concern with some types of engineered wood products. *Problem*: Engineered wood subsystems. *Activities*: Formulate technology for characterizing the structural performance of light-frame subsystems in terms of the properties of the lumber, panel products, and connections used in their construction. *Problem*: Engineered building systems.	*Problem*: Properties of engineered wood products. *Activities*: Characterize the properties of engineered wood products for safe, efficient design of structures, and efficient use of the forest resource by determining the suitability of various materials and designs and their performance characteristics in engineered wood products. *Problem*: Structural performance of wood-building systems. *Activities*: Create new technologies, guidelines, and knowledge that improve the quality, safety, durability and reliability of buildings, particularly housing. *Problem*: Moisture control in buildings. *Activities*: Create new technologies and guidelines that minimize the potential for moisture damage to the structure while maintaining the energy efficiency of the building and indoor air quality for the occupants.	*Problem*: Building moisture design methodology. *Activities*: Develop a performance-based moisture design approach for wood frame buildings that is firmly based on technical information and engineering principles. *Problem*: Wetting and drying of wood-frame buildings and building components. *Activities*: Provide information to better predict the potential for mold and decay in wood building components. *Problem*: Moisture and thermal properties of wood products. *Activities*: Build a database for wood products, with a special emphasis on wood composites. Make the Forest Products Laboratory a center for collecting and documenting reliable, unbiased property data for use in hygrothermal models and building design.

Table 14. (Continued)

Forest Products Laboratory's research work unit and plan components	Research work unit title, and period and activities covered		
	Activities: Relate the performance of major components—floors, walls, and roof—to the performance of the entire structure. Develop computer programs that can be used by engineers. *Problem*: Engineered timber bridge systems. *Activities*: Develop new technology for engineered timber bridge systems to provide economical alternatives to bridges of other materials and assist in implementing the technology in national design standards. Problem: In-place evaluation of structures. Activities: Improved methods are needed to determine the in-place capacity of existing structures in order to extend their life. Problem: Moisture management in buildings. Activities: Develop better criteria for the design and operation of energy-efficient wood buildings for satisfactory performance with regard to moisture.	*Problem*: Performance of wood transportation structures. *Activities*: Improve existing wood transportation structures and develop new systems that conserve and improve wood use, and improve the adequacy and condition of the U.S. transportation infrastructure.	*Problem*: Durability of composite wood products. *Activities*: Industry needs assistance in the development of better industrywide standards and testing methods (for hardboard and oriented strandboard (OSB) siding, and more recently, with OSB sheathing) that would provide a more consistent product with greater durability.

Table 14. (Continued)

Forest Products Laboratory's research work unit and plan components	Research work unit title, and period and activities covered		
Research work unit and plan components			
4719			
	Wood Processing and Drying Systems, 1993-98	Wood Engineering and Drying Systems Design Criteria, 1998-2004	Condition Assessment and Rehabilitation of Structures, 2004-05
Area of research applicability	National	National	National
Mission	Develop primary wood processing systems that have minimum impact on our environment and enhance the value and yield of products obtained from forest resources.	Develop nondestructive evaluation technologies, engineering design criteria, and drying systems that facilitate use of underutilized materials and help conserve the forest resource.	Develop nondestructive evaluation technologies, structural analysis procedures, inspection methods, and rehabilitation procedures for wood structures.
Research problem and activities	*Problem*: Need to develop analytical tools to define product recovery. *Activities*: Development of a series of computer models of key processing subsystems. These models will be developed by using existing processes and quality control information supplemented with mill product yield information. *Problem*: A lack of understanding of the dynamic variables of sawing hinders improvements in sawing technology.	*Problem*: Need to develop drying technologies for economical use of the changing resource base while reducing energy costs and environmental concerns. *Activities*: Work toward developing new strategies by, among other things, estimating changed or previously unknown drying characteristics and developing fundamental heat and mass transfer information with an eye toward computer simulations of kiln drying. *Problem*: Need to develop engineering design analysis methods and information for designing efficient wood structures.	*Problem*: Need to develop baseline analysis methods and information for designing and inspecting wood structures. *Activities*: Establish baseline analysis methods for design and inspection of wood structures. Research will be applicable to building codes to allow for new wood-based materials, advanced connections, and inspection efficiencies. *Problem*: Need to develop and evaluate new nondestructive technologies for assigning engineering properties to wood products and structural systems.

Table 14. (Continued)

Forest Products Laboratory's research work unit and plan components	Research work unit title, and period and activities covered		
	Activities: Basic research aimed at developing an understanding of the primary dynamic variables in sawing, and the means to control them. *Problem*: There is a need for improved technology and techniques for manufacturing and drying quality veneer. *Activities*: The Modular Veneer Press dryer has demonstrated the capability of reducing buckling when drying fine hardwood veneers. Final engineering design efforts will lead into commercial development. *Problem*: There is a need for better drying techniques, appropriate pre-sorting technology, and tools to better predict and control the response of lumber to kiln drying.	*Activities*: Establish methods and data for designing efficient and reliable wood structures. Applicable to building codes to (1) ensure adequate design for high winds and earthquakes, and (2) allow for new wood-based materials and design efficiencies. *Problem*: Need to develop and evaluate new nondestructive technologies for assigning engineering properties. *Activities*: Develop nondestructive techniques that provide information regarding the properties of wood products used in a wide variety of applications and under a wide range of environmental conditions. Problem: Need to develop new design procedures for repairing components, assemblies, and systems.	*Activities*: Develop nondestructive evaluation and assessment technologies and to understand the influence of environmental variables on their reliability. *Problem*: Need to develop new procedures for in-place assessment and new methods of field repair to extend service life of timber bridges. *Activities*: Despite technological gains for the inspection of other materials, inspection methods and equipment for timber transportation structures have remained virtually unchanged for the past 50 years. *Problem*: Need to develop new design procedures for repairing components, assemblies, and systems. *Activities*: Provide fundamental knowledge of how to assess the condition of a structure, how the decay and defects affect the strength and reliability of the structure, and how to repair the structure.

Table 14. (Continued)

Forest Products Laboratory's research work unit and plan components	Research work unit title, and period and activities covered		
	Activities: Develop physical property data and analytical methods to predict the response of wood to drying in terms of drying time, moisture content gradients, and stress; and establish relationships between these responses and relevant material and process variables. *Problem*: There is a need for better information on environmental aspects of wood drying and environmentally appropriate drying technology. *Activities*: Develop and transfer information and readily applied technology to promote environmentally responsible lumber-drying technologies.	*Activities*: Provide fundamental knowledge of how to assess the condition of a structure, how the decay and defects affect the strength and reliability of the structure, and how to repair the structure.	
4722	Modification of Lignocellulosics for Advanced Materials and New Uses, 1993-97	Modified Lignocellulosic Materials, 1997-2002	Modified Lignocellulosic Materials, 2002-07
Area of research applicability	National	National	National
Mission	To develop advanced paper-based products and composite materials based on sustainable property enhanced bio-fibers alone or in combination with other resources.	To develop advanced environmentally friendly composite materials from chemically and physically modified wood-based resources alone or in combination with other resources to extend the use of our forest resources.	To develop advanced environmentally friendly composite materials from chemically and physically modified wood-based resources alone or in combination with other materials to extend the use of our forest resources.

Table 14. (Continued)

Forest Products Laboratory's research work unit and plan components	Research work unit title, and period and activities covered		
Research problem and activities	*Problem*: Need to understand chemical and physical properties of lignocellulosics to identify new sources of polymeric feedstocks and advanced composite materials. *Activities*: Determining chemical and physical properties of the wide varieties of biobased fibers that can be used in combination with wood-based fiber for composite materials. *Problem*: Need to understand chemical and morphological modifications of lignocellulosics to improve properties and to maximize end-use performance of paper-based products and biobased materials.	*Problem*: Need to understand the relationship between the chemical and physical properties of wood-based materials and final composite properties to identify new sources of polymeric feedstocks and advanced composite materials. *Activities*: Determining chemical and physical properties of the wide varieties of lignocellulosic-based fibers that can be used in combination with wood fiber for composite materials. *Problem*: Need chemical and structural modifications of wood-based materials to improve properties and to maximize end-use performance of biobased composite materials. *Activities*: Chemically modify lignocellulosic resources to enhance properties such as dimensional stability and to characterize the resistance of wood composites after modification when they are exposed to adverse environments.	*Problem*: Need to understand the relationship between the chemical and physical properties of wood-based materials and final composite properties to identify new sources of polymeric feedstocks and advanced composite materials. *Activities*: Determine chemical and physical properties of the wide varieties of lignocellulosic-based fibers that can be used in combination with other materials to produce new generations of composite materials. *Problem*: Need for chemical and structural modifications of wood-based materials to improve properties and to maximize end-use performance of biobased composite materials.

Table 14. (Continued)

Forest Products Laboratory's research work unit and plan components	Research work unit title, and period and activities covered		
	Activities: Chemically modify biobased resources to enhance properties such as dimensional stability and to characterize the resistance of wood composites after modification when they are exposed to adverse environments. *Problem*: Improved methodologies are needed for the development of ligocellulosic and nonlignocellulosic composites by understanding the materials science and the matrix/interface morphology of these resources. *Activities*: Provide data for industry to produce a whole new line of value-added products taking advantage of the properties of many different types of resources.	*Problem*: Improved techniques are needed for better understanding of the interface and interphase chemistry between wood-based resources and other resources to improve surface interactions. *Activities*: Provide data for industry to produce a whole new line of value-added products taking advantages of the properties of many different types of resources. *Problem*: Environmentally friendly technologies are lacking for the development of new wood-based composite materials that will lead to more efficient use of our forest resources. *Activities*: Develop technology that results in environmentally friendly lignocellulosic composite materials.	*Activities*: Chemically modify lignocellulosic resources to enhance properties such as water repellency, and to characterize the resistance of wood composites after modification when they are exposed to adverse environments. *Problem*: Improved techniques are needed to convert forest biomass into geotextiles and filters and to maximize effectiveness in soil stabilization and removal of contaminates from water. *Activities*: Use small-diameter trees, waste biomass from our forests and agricultural land, break them down into usable particles and fibers, and form geotextiles and filters. *Problem*: Improved technologies are needed to understand nature's chemical and biological degradation mechanisms and to interfere with these degradation processes to extend the useful life of a biobased composite. *Activities*: Develop an understanding of the chemistries involved in the degradation of wood and use that information to develop new durable products that are environmentally friendly.

Table 14. (Continued)

Forest Products Laboratory's research work unit and plan components	Research work unit title, and period and activities covered		
4723	Wood Preservation, 1994-97	Wood Preservation and Fire Research, 1997-2002	Wood Preservation and Fire Safety Engineering, 2002-05
Area of research applicability	National	National	National
Mission	To enhance the durability of wood through the development of environmentally innocuous products and technologies.	To develop fire safety and wood preservation data, methodologies, and technologies that improve or facilitate value-added utilization and recycling of wood products.	To improve the durability and fire safety of forest products in the context of changing environmental and societal needs, thereby sustaining forests, the economy, and the quality of life
Research problem and activities	*Problem*: Need to develop innovative, environmentally preferable preservative systems for the production of durable wood products. *Activities*: Identify and develop preservative treatments to provide wood products that meet multiple performance requirements (such as durability) while also satisfying such requirements as environmental safety, ease of disposal, and recycling. *Problem*: Need to develop new methodologies for abbreviating the time necessary to perform evaluations of the long-term performance of wood preservatives.	*Problem*: Need to develop new environmentally preferable preservative systems for durable wood products. *Activities*: Identify and develop treatments to provide wood products that meet multiple performance requirements such as durability, while also satisfying such requirements as environmental safety, ease of disposal, and recycling. *Problem*: Need to develop methodologies for the accurate prediction of long-term performance of durable wood species and treatments.	*Problem*: Need to develop the data and models for fire safety engineering of forest products in a performance-based building code environment. *Activities*: Obtain property and validation data and simulate full-scale fire scenarios with computer fire models. Primary emphasis will be the growth of a fire from ignition to "flashover." *Problem*: Need to document and optimize fire safety of engineered wood products in structural applications.

Table 14. (Continued)

Forest Products Laboratory's research work unit and plan components	Research work unit title, and period and activities covered		
	Activities: Develop methodologies that will permit the rapid and quantitative evaluation of long-term preservation performance. *Problem*: Lack of adequate knowledge of preservative processing technologies for producing durable products from our changing forest resources. *Activities*: Research will address solid and composite wood products and will include investigations of treatments for softwood and hardwood species that are currently underutilized.	*Activities*: Develop methodologies that permit rapid and quantitative evaluation of long-term preservative performance. *Problem*: Need to improve the treatability and durability of underutilized species and new wood composites and the recycling of existing treated materials. *Activities*: Research will address solid and composite wood products and will include investigations of treatments for softwood and hardwood species that are currently underutilized. *Problem*: Need to develop information and methodologies that will improve the service life of wood structures in the wildland-urban interface. *Activities*: Research efforts will emphasize the responsibility of homeowners and supply guidelines concerning the structure and immediate surrounding areas.	*Activities*: Obtain data on the components of the engineered wood products, develop models to better understand their fire performance, and develop recommendations on how to optimize their fire performance. *Problem*: Need to improve the survivability of wood structures in the wildland-urban interface. *Activities*: Reduce the fire hazards in the wildland-urban interface by improving the quality and flexibility of fire safety recommendations for wood structures. Such recommendations are part of the National Fire Plan community assistance programs. *Problem*: Improved methods of evaluation and analysis of test samples and data are needed for timely and accurate prediction of long-term performance of durable forest products.

Table 14. (Continued)

Forest Products Laboratory's research work unit and plan components	Research work unit title, and period and activities covered		
		Problem: Need to develop fire safety data and methodologies to preserve or expand the use of wood products that conserve our wood resources and use new timber resources. *Activities*: Research emphasis will be on the ability of wood products to contribute to fire growth.	*Activities*: Methodologies for evaluating the durability of forest products include the soil bottle test, the larger-scale fungal cellar, and the field plots. We anticipate assessing potential improvements to all these levels of evaluations. *Problem*: Need to reduce environmental impact of treated forest products. This problem area will have two main elements: new, more environmentally compatible treatments for wood protection and assessment, and reduction of preservative chemicals released into the environment. *Activities*: Investigate various methods for reducing the levels of preservative chemicals in the wood. *Problem*: Need to improve ability to treat a diverse range of species and underutilized small-diameter materials for treated wood applications. *Activities*: Efforts to evaluate both conventional and alterative preservatives in a wide range of hardwood and softwood species will continue.

Table 14. (Continued)

Forest Products Laboratory's research work unit and plan components	Research work unit title, and period and activities covered	
4724	Statistical Methods in Wood and Fiber Research, 1998-2004	Statistical Methods in Wood and Fiber Research, 2004-05
Area of research applicability	National	National
Mission	To enhance the integrity and efficiency of the Forest Products Laboratory's research efforts through the development, evaluation, and promotion of modern statistical methods.	To enhance the integrity and efficiency of the Forest Products Laboratory's research efforts through the development, evaluation, and promotion of modern statistical methods.
Research problem and activities	*Problem*: Need to enhance the quality of wood utilization research and economic assessments. *Activities*: Enhancing the Forest Products Laboratory's research through collaborative research with other laboratory scientists, professional support to the scientists, and the transfer of research-derived technology in the form of user-friendly computer programs that provide new capabilities to the scientists. *Problem*: Need to improve the statistical modeling of properties, processing, and the performance of wood, fiber, and composites. *Activities*: Develop improved statistical modeling methodology needed by the Forest Products Laboratory's research programs.	*Problem*: Need to enhance the quality of wood utilization research and economic assessments. *Activities*: Enhancing the Forest Products Laboratory's research through collaborative research with other laboratory scientists, professional support to the scientists, and the transfer of research-derived technology in the form of user-friendly computer programs that provide new capabilities to the scientists. *Problem*: Need to improve statistical modeling of properties, processing and performance of wood, fiber, and composites. *Activities*: Research efforts focus on developing the improved statistical modeling methodology needed by the Forest Products Laboratory's research programs and the evaluation of existing statistical methods.

Table 14. (Continued)

Forest Products Laboratory's research work unit and plan components	Research work unit title, and period and activities covered	
4725 (new research work unit Dec. 2003)		Fire Safety, 2003-04
Area of research applicability		National
Mission		To develop data, methodologies, and technologies needed to ensure that wood products and wood-based structures do not adversely contribute to the loss of life and property in fires.
Research problem and activities		*Problem*: Data and models are required for fire safety engineering of forest products in a performance-based building code environment. *Activities*: Obtain property and validation and simulate full-scale fire scenarios with computer models. Primary emphasis will be the growth of a fire from ignition to a flashover. *Problem*: Need to document and optimize fire safety of engineered wood products in structural applications. *Activities*: Obtain data on the components of the engineered wood products, develop models to better understand their fire performance, and develop recommendations on how to optimize their fire performance. *Problem*: Need to improve survivability of wood structures in the wildland-urban interface. *Activities*: Developing improved safety recommendations will include databases of material properties, investigations of potential fire scenarios, and identification of design changes to reduce hazards.

Table 14. (Continued)

Forest Products Laboratory's research work unit and plan components	Research work unit title, and period and activities covered		
4851	Timber Demand and Technology Assessment Research, 1994-97	Timber Demand and Technology Assessment Research, 1997-2002	Timber Demand and Technology Assessment Research, 2002-04
Area of research applicability	National	National	National
Mission	To develop long-term projections of consumption, production, price, manufacturing costs, and conversion yields for pulp and paper, fuel wood, hardwood lumber, and nonstructural panel markets; and manufacturing cost and conversion yields for softwood lumber and structural panel markets; to assess new technologies for converting timber into new or improved paper and wood products.	To provide economic information, analysis, and projections indicating (1) how wood is used in the economy, (2) how and why wood use changes over time, (3) changes in the kinds and amounts of wood and fiber needs, (4) natural resources management needs, (5) market equilibrium quantities and prices, and (6) selected environmental impacts.	To provide economic information, analysis, and projections indicating how and why the markets and technologies for wood products change over time, implications for natural resources management, and selected broad environmental and social impacts.
Research problem and activities	*Problem*: Improvements in the Forest Products Laboratory's Pulpwood Model are needed to provide long-term projections of production, consumption, price, costs, and conversion yields for the pulp and paper market. *Activities*: Research will provide long-term projections of regional consumption, production, and price in North America for pulpwood and recycled fiber, wood, pulp, and paper and board.	*Problem*: There is a lack of understanding of the effects of long-range market equilibria, production, and technological changes on the pulp and paper industry. *Activities*: Improving modeling and analysis with the North American Pulp and Paper model for the Forest and Rangeland Renewable Resources Planning Act of 1974 (RPA) assessments requires a range of research studies and collaboration with research partners.	*Problem*: There is a need to monitor market trends and improve the understanding and modeling of market equilibria, production trends, and technology changes of the U.S. pulp and paper industry. *Activities*: Economic modeling and analysis of market trends in the U.S. pulp and paper sector for RPA assessments and other purposes requires a range of research studies and collaboration with research partners.

Table 14. (Continued)

Forest Products Laboratory's research work unit and plan components	Research work unit title, and period and activities covered		
	Problem: New economic models are needed to provide long-term projections of production, consumption, price, installation costs, and conversion yields for residential and industrial wood energy markets. *Activities*: Economic models will be developed to produce long-term projections of Installation costs and factors for converting fuel wood to energy. *Problem*: Economic models are needed to provide long-term projections of production, consumption, price, costs, and coercion yields/for lumber and panel markets. *Activities*: Provide long-term projections on regional consumption in North America for timber consumed in the manufacture of hardwood lumber and nonstructural panels.	*Problem*: There is a lack of understanding of long-range production and technological changes in the solid wood industries and solid wood end-use industries. *Activities*: Provide a model of the solid wood sector to project interaction of end-use demand, technology change, and supply sources to determine wood and fiber requirements. *Problem*: There is a lack of readily available, consistent information on timber, wood products, and paper products industries for forestry and forest industry strategic planning. *Activities*: Maintain national data series using most appropriate sources, and develop new data series, in cooperation with users, to show new industry developments and adjust series as needed.	*Problem*: Improved understanding is required about the effect of production trends and technology changes in solid wood industries and end-use industries on product and wood needs. *Activities*: Develop estimates of the amounts of solid wood waste available in the United States, and assess opportunities to increase its utilization in recycled products. *Problem*: Readily available, consistent information is needed on timber, wood products, and paper products industries for forestry and forest industry strategic planning. *Activities*: Maintain national data series using most appropriate sources and develop new data series, in cooperation with users, to show new industry developments and adjust series as needed.

Table 14. (Continued)

Forest Products Laboratory's research work unit and plan components	Research work unit title, and period and activities covered		
	Problem: Economic analysis is needed to assess new technologies for converting timber into new or improved paper and wood products. *Activities*: Assess new technologies for producing new or improved paper and wood products. The technologies evaluated will be those developed at the Forest Products Laboratory.	*Problem*: There is a lack of understanding of the effects of sustainable forestry efforts and certification of forests and forest products on demand for and supply of wood in the United States. *Activities*: Provide information to aid in projecting market trends and technology change in the pulp and paper industry and the solid wood industry.	*Problem*: Knowledge is needed on the economic viability of new technologies for converting wood and fiber into new or improved paper and wood products. *Activities*: Evaluate market prospects and material needs for alternative products that would use small-diameter timber in the West.

Source: Forest Service documents.

Note: The Forest Products Laboratory's research work unit activities for 4708, 4715, and 4718 are not listed in the table because they were combined with other units during the 1990s.

Table 15. Conducted and Planned Activities for the Forest Service's Research Work Units Associated with Research Stations

Research station— associated work unit, and plan components	Research work unit title, and period and activities covered		
Northeastern Research Station	Improved Processing Technology for Hardwoods, 1990-97	Efficient Use of the Northern Forest Resource, 1997-2003	Efficient Use of the Northern Forest Resource, 2003-08
4701			
Area of research applicability	Eastern U.S. and wherever hardwood lumber and products are processed.	Northern U.S. and hardwood growth and processing regions nationwide.	Northern U.S. and hardwood growth and processing regions nationwide.
Mission	To increase effective use of hardwood resources through research to improve processing technology.	To promote natural resource conservation and help sustain forest-based industries and economies through research and development on resource characteristics, forest management, and innovative processing technologies and their influence on conversion efficiency in solid wood processing.	To develop and deliver knowledge and innovative technology that improves efficiency in forest products conversions to strengthen U.S. worker productivity in global wood products marketplace; increase the value of the timber resource and the economic viability of forest management options.
Research problem and activities	*Problem*: Techniques are needed to evaluate production and economic implications of using new processing technologies. *Activities*: Develop realistic simulation models to evaluate technical characteristics of integrated hardwood processing systems using new technology.	*Problem*: Research is needed to determine how wood quality and utilization opportunities are affected by silvicultural systems and forest operations. *Activities*: Determine the effects of silvicultural systems and forest operations on wood quality (including physical and mechanical properties, decay, and bacterial infection) of certain northern hardwood species.	*Problem*: Research is needed to determine how wood quality and utilization can be enhanced through silvicultural and forest pest management systems and operations. *Activities*: Determine the effects of silvicultural systems, forest pests, and forest operations on wood quality of selected eastern hardwood species.

Table 15. (Continued)

Research station— associated work unit, and plan components	Research work unit title, and period and activities covered		
	Problem: Standards and test criteria are needed for evaluating computer-numeric controlled (CNC) machinery and associated tooling. *Activities*: Develop standards for evaluating the accuracy, efficiency, and safety of CNC woodworking machinery. *Problem*: More efficient techniques are needed to make existing information available to decision makers in a timely manner and in a form suitable to answer specific questions. *Activities*: Quantify the existing knowledge in specific areas of wood processing and develop expert systems using this information.	*Problem*: More effective production control methods and software tools must be developed to analyze the impact of current and proposed production systems on efficiency of resource utilization. *Activities*: Develop optimal rough-mill processing strategies and product recovery benchmarks for different lumber characteristics through research into interactions among such factors as lumber size and grade, and develop improved lumber-scanning technology.	*Problem*: More effective manufacturing strategies and production control technologies must be developed and implemented so that solid wood products manufacturers can improve resource utilization efficiency and international competitiveness. *Activities*: Develop and enhance the adoption of optimal rough-mill processing and product-recovery strategies by researching interactions among such factors as lumber size and grade.
4751	Forest Engineering Research Systems to Integrate Harvesting with Other Resource Management Objectives, 1988-97	Forest Engineering Research-Systems Analysis to Evaluate Alternative Harvesting Strategies, 1997-2003	Integration of Forest Operations Into Eastern Hardwood Intermediate Cuttings and Structural Retention Treatments, 2003-08
Area of research applicability	Mountainous, ridge and valley, and plateau regions of the eastern United States.	United States	Northeastern United States

Table 15. (Continued)

Research station— associated work unit, and plan components	Research work unit title, and period and activities covered		
Mission	To use systems analysis to effectively integrate silviculture, logging technology, economics, and wildlife management into a forest management decision model with particular applications to steep terrain.	To use systems analysis to evaluate alternative harvesting strategies for all forested geographic regions in the Northeast.	To improve and integrate forest operations to accomplish intermediate silvicultural treatments in hardwood forests of the Northeast.
Research problem and activities	*Problem*: Forest management planning tools are inadequate for steep terrain. *Activities*: The goal of this research is to develop a system analysis computer model, MANAGE, that adequately models the biological, technological, economical, and silvicultural treatment of a forest stand over a rotation. *Problem*: Methods are inadequate to plan and carry out multiproduct harvesting and utilization of eastern hardwoods on steep terrain. *Activities*: Systems theory would be used to model multiproduct harvesting and utilization of eastern hardwoods on steep terrain in order to measure the impact of multiproduct harvesting on forest management.	*Problem*: Management planning models are inadequate for forest operations and forest product transportation. *Activities*: The goal of this research is to develop a systems analysis computer model that adequately incorporates validated growth models, logging technology, wildlife, economical, and silvicultural treatments of a forest stand over a rotation. *Problem*: There is a lack of complete synthesis on all aspects of the forest operations process, selection of the harvesting process, multiproduct harvesting, loss caused by log damage, and other procedures.	*Problem:* There is a lack of synthesis of the forest operations process to improve intermediate cuttings and structural retention harvests for eastern hardwoods. *Activities*: The goal of this research is to provide better information and synthesis of the forest operations process so that economical decisions can be made when selecting and implementing intermediate cuttings and structural retention treatments.

Table 15. (Continued)

Research station— associated work unit, and plan components	Research work unit title, and period and activities covered		
		Activities: The goal of this research is to provide better information and synthesis of all phases of the forest operations so that better management models can be developed. Information will be synthesized with other stand data to determine the maximum revenue that can be expected from a stand.	
4803			
	Analysis of Domestic and International Hardwood Product Markets, 1992-97	Economics of Eastern Forest Use, 1997-2003	Eastern Forest Use in a Global Economy, 2003-08
Area of research applicability	United States and countries with which it conducts trade in hardwood products	National, but with a primary focus on hardwood and softwood forest resources in the eastern United States.	National, but with an emphasis on issues affecting the hardwood forest industry and resources of the eastern United States.
Mission	To develop economic and technical information which enables concerned individuals and organizations to make more informed decisions regarding the use of forest resources and the production and marketing of forest products.	To develop annual estimates of primary and secondary hardwood product production and consumption, identify and analyze economic and other factors or events that are likely to significantly alter wood production and consumption patterns and future trends, and examine and analyze alternatives (opportunities) for extending and conserving the hardwood resource and improving the efficacy of hardwood forest management, utilization, and rural development initiatives.	To provide economic, market, and wood-use information that will support the health and sustainability of forest-based industries, hardwood forests, and forest communities in the eastern United States.

Table 15. (Continued)

Research station— associated work unit, and plan components	Research work unit title, and period and activities covered		
		Problem: Information is needed on all markets for roundwood and on the principal consumers (domestic and overseas) of hardwood lumber and wood fiber on a continuing basis.	
Research problem and activities	*Problem*: There is a need to explain the interrelationships among domestic markets for hardwood products. *Activities*: Develop refined databases for hardwood product production, usage, and price. *Problem*: There is a need to develop more detailed information on international hardwood products and describe the impact of international trade on domestic hardwood product markets. *Activities*: Collect and validate all domestic and foreign data relating to the exportation or importation of hardwood products from or into the United States.	*Activities*: Compile statistics on primary hardwood product production in cooperation with FIA and assume responsibility within the Northeastern Research Station for activities associated with the TPO. *Problem*: Research is needed on viable, market-based, value-added product opportunities that improve use of the forest resource, enhance achievement of resource management objectives, and sustain rural communities. *Activities*: Research conducted in this area will be product, process, site, and market-specific.	*Problem*: There is a need to benchmark and monitor hardwood use to assess the impacts of change and competing demands on the forest industry, forest resource, and forest-based communities of the eastern United States. *Activities*: Compile statistics on roundwood receipts by primary processors within the 13 states comprising the Northeastern Research Station. *Problem*: Need to provide information, analysis, and tools to assist in assessing competitive environments and evaluating product, market, and management opportunities and challenges facing U.S. producers of wood and wood-based products so as to more effectively manage and utilize the U.S. eastern hardwood resource.

Table 15. (Continued)

Research station— associated work unit, and plan components	Research work unit title, and period and activities covered		
	Problem: There is a need to develop market-based decision-making tools that can be used by hardwood processors (secondary and primary) and rural development personnel. *Activities*: Develop a profile of common characteristics based on different types of primary and secondary hardwood manufacturers on the basis of size, productive inputs purchased, marketing practices, and product(s) produced.		*Activities*: Studies will be conducted to determine the effects of size, product type, management style and quality, financial structure, employee relations, material cost, and other factors that affect competitiveness.
4805	Enhancing the Performance and Competitiveness of the U.S. Hardwood Industry, 1994-2000	The Influence of Markets on the Sustainability of Eastern Hardwood Forests, 2000-05	
Area of research applicability	Eastern United States	Primary focus on the hardwood region of the eastern United States	
Mission	To explore alternative strategies that will improve the long-term performance and competitiveness of the U.S. hardwood products industry and analyze the impact of these strategies on the hardwood resources.	To examine interrelationships between forest product markets and the composition, structure, and sustainability of the eastern hardwood forest.	

Table 15. (Continued)

Research station— associated work unit, and plan components	Research work unit title, and period and activities covered		
Research problem and activities	*Problem*: There is a need to analyze the structure, conduct, and performance of the various hardwood products industries that utilize hardwood, roundwood, or hardwood timber. *Activities*: Classify, in economic terms, the structure of the major primary hardwood processing industry on a regional basis and assess the conduct and performance of each industry based on this classification. *Problem*: There is a need to assess the impact of alternative intervention approaches to remedy externalities from the production, harvesting, and processing of hardwood timber. *Activities*: Examine intervention mechanisms that are currently being used to remedy market externalities associated with hardwood timber production, harvesting, and processing.	*Problem*: There is a need to examine current impact of different localized market situations on forests and provide information that will allow policy makers, industry, and others to assess forest sustainability under current market-induced disturbance regimes. *Activities*: Develop a framework to examine market-induced disturbance utilizing the constructs of economic theory in combination with current and historic information on the changing forest industry and markets. *Problem*: There is a need to develop procedures to predict how future market-induced disturbance will change species distribution, timber quality, and the sustainability of specific ecosystems and examine how adaptive technology will influence these changes. *Activities*: Relate long-term changes in price, demand, and biological supply to changes in technology and the type of forest that will be impacted by the technology.	

Table 15. (Continued)

Research station— associated work unit, and plan components	Research work unit title, and period and activities covered		
	Problem: There is a need to explore strategies that will help hardwood processing firms and industries remain competitive while adhering to the mandates of a changing society. *Activities*: Isolate the major "unresolvable issues" that are likely to polarize environmental organizations and the hardwood industry and examine the major reasons why these conflicts exist.		
Pacific Northwest Research Station			
4865			
	Human and Natural Resources Interactions Research Program, 1995-2002	Human and Natural Resources Interactions Research Program, 2002-current	
Area of research applicability	No information.	Utilization, economics, and social science research; regional (Pacific Northwest), national, and international scales.	
Mission	No information.	To improve understanding of social and economic values as input to and evaluation of resource management decisions.	
Research problem and activities	*Problem*: Ecologically sustainable production of forest activities.		

Table 15. (Continued)

Research station— associated work unit, and plan components	Research work unit title, and period and activities covered		
	Activities: Evaluate the influence of specific stand structure-manipulation operations on the quality and quantity of forest products.	*Problem*: There is a need to improve knowledge of the opportunities to use local, regional, and international markets and how to use management practices to sustain diversity of forest conditions and outputs. *Activities*: Improve understanding of how the forest sector (broadly defined) functions, including intrasectoral dynamics, and interactions between the forest sector and other sectors.	
Pacific Southwest Research Station			
4202			
	No wood utilization	No wood utilization	Sierra Nevada Research Center, 2005-15
Area of research applicability			Local, regional, and international
Mission			Sierra Nevada ecosystems are complex and our knowledge of them is incomplete. As a result, the long-term outcome of any given land and resource management strategy is uncertain. We will provide assistance to land managers and policy makers by addressing this management dilemma through targeted research emphasizing an integrated, ecoregional approach to examine particular physical, ecological, and socio-economic issues, across a range of appropriate spatial and temporal scales specific to each issue.

Table 15. (Continued)

Research station— associated work unit, and plan components	Research work unit title, and period and activities covered		
Research problem and activities			*Problem*: Institutional and policy processes. *Activities*: Develop appropriate research methodologies to understand institutional processes by which resource values are established as public goods.
Rocky Mountain Research Station			
4156			
	No wood utilization	No wood utilization	Southwestern Forest Health Restoration and Wildland-Urban Interface Fuels Management, 2005-
Area of research applicability			Unknown
Mission			Unknown
Research problem and activities			*Problem*: A need exists to develop and understand economics, markets, and utilization opportunities to support the management and restoration of southwestern forests that also contribute to the economic vitality of local and regional communities. *Activities*: Understand the economic and social impacts of alternative forest management options on communities and sustainable utilization opportunities.

Table 15. (Continued)

Research station— associated work unit, and plan components	Research work unit title, and period and activities covered		
Southern Research Station			
Area of research applicability	Southern United States	Southeastern United States	Southern United States
4104 (previously 4104 and 4701)	Utilization of Southern Timber (4701), 1993-98	Ecology and Genetics of Southern Pine Ecosystems, 1994-99 (4104)	Disturbance and the Management of Southern Pine Ecosystems, 1999-2004
Mission	To define how environmental and socioeconomic factors will change the southern timber resource; and develop fundamental information and technology to utilize this resource while conserving biodiversity and providing for a sustained supply of raw material.	To develop scientific knowledge needed for regenerating, managing, protecting and improving pine plantations, natural stands, and ecosystems.	To increase understanding and develop applications of disturbance to sustain the productivity and functions of southern pine ecosystems.
Research problem and activities	*Problem*: Information is lacking on the effects of site, species composition, stockings, rotation age and their interactions on value, yield, anatomical, physical and mechanical properties of southern timber, which is a barrier to meeting management and utilization goals. *Activities*: Three areas of research emphasis: (1) value and yield of forest products; (2) anatomical and physical properties; and (3) mechanical properties. These areas are interrelated.		*Problem*: There is a need to better understand the effects of within tree, stand, environmental, and forest management factors on wood properties to meet productivity and sustainability objectives. *Activities*: The emphasis of this research is on relating basic wood properties to and within its tree, stand, environmental, and management factors that affect southern pine wood formation.
4701	Utilization of Southern Forest Resources, 1993-99	Utilization of Southern Forest Resources, 1999-2005	Utilization of Southern Forest Resources, 2005-10

Table 15. (Continued)

Research station— associated work unit, and plan components	Research work unit title, and period and activities covered		
Area of research applicability	Southern United States and tropical areas	Southern United States and tropical areas	Southern United States and tropical areas
Mission	To define and apply fundamental chemistry, material science, and engineering principles to the utilization and processing of southern forest resources in an environmentally sound way.	To define and describe the fundamental raw material characteristics influencing the sustainable and environmentally sound use of southern forest resources.	To define and describe the fundamental raw material characteristics influencing the sustainable and environmentally sound use of southern forest resources.
Research problem and activities	*Problem*: Little is known of the composition of secondary metabolites in plants of forest communities, their biological significance, or their potential value as a renewable source of specialty chemicals. *Activities*: Define the chemistry and significance of secondary metabolites of selected plants in the forest community to help determine the biological significance of these compounds. *Problem*: Information is inadequate on the physical and mechanical properties of primary wood constituents to analytically model and understand the properties of many wood-based composites.	*Problem*: A gap in our understanding of the fundamental physical properties that are central to the biological/ecological significance of plant polyphenols or terpenes limits their usefulness. *Activities*: Define fundamental physical properties that are important in determining the biological or ecological significance of secondary plant metabolites. *Problem*: There is a need to improve composite properties by evaluating the physical and mechanical properties of primary wood constituents.	*Problem*: Information is insufficient on the physical, chemical, and mechanical properties of woody raw material originating from southern forests. *Activities*: This problem area will focus on three elements pertinent to the effective utilization of our southern forest resource. *Problem*: There is a need to better design biobased products based on the properties and composition of the southern forest raw material. *Activities*: Research on this problem area will be accomplished by a three-pronged approach, which includes the relationship between component properties and their composite products.

Table 15. (Continued)

Research station— associated work unit, and plan components	Research work unit title, and period and activities covered		
	Activities: Research will focus on development of new structural wood-based composites as well as improvements on conventional composites. *Problem*: New adhesive systems and bonding processes are needed to enhance adhesion, reduce bonding costs, and provide technological improvements. *Activities*: This research will focus on defining the performance of adhesive bonding process for wood-and-polymer composites and developing new processing technology to utilize mixture of recycled wood fiber and plastics for high-value products. *Problem*: More understanding of the fundamental chemistry and physics of interaction between synthetic polymers and wood materials is needed to accommodate the use of wood in new composite systems.	*Activities*: Research will focus on property assessment and material characterization as they relate to the structural performance of wood-based composites. *Problem*: There is a lack of adequate understanding of interfacial structure and properties to develop new composite material development. *Activities*: This problem area will study the relationship between wood surface properties and interfacial characteristics. *Problem*: There is a need to improve effective utilization of wood from difficult-torecycle and intensively managed sources into high performance composite products. *Activities*: This research will support the development of intensively-cultured production approaches for wood by investigating this material's performance in existing process technologies and composite systems.	

Table 15. (Continued)

Research station— associated work unit, and plan components	Research work unit title, and period and activities covered		
	Activities: The research in this area will continue along the program lines that have been established over the last several years of investigation that address the physicochemical interaction between wood fibers and synthetic polymers.	*Problem*: There is a need to sufficiently understand the relationships between tree growth variables, fundamental wood properties, and end-product performance for optimal utilization of the forest resource. *Activities*: This research area will focus on the development and use of rapid assessment methods to study wood characteristic/product property relationships at the molecular level.	
4702	Primary Hardwood Processing and Products, 1988-97	Integrated Life Cycle of Wood: Tree Quality, Processing, and Recycling, 1997-2003	Forest Products Conservation, 2003-08
Area of research applicability	National	National	National
Mission	To identify, evaluate, and develop new or improved automated primary hardwood processing technologies and hardwood products that make U.S. industry more competitive in domestic and foreign markets.	To enhance wood resource conservation and sustainability through advanced timber analysis and wood processing, and effective wood product recovery, reuse, and recycling.	To enhance sustainable forest resource use through improved product, processing, and recycling analysis.
Research problem and activities	*Problem*: There is a need to develop improved and new automated primary hardwood processing technologies.	*Problem*: There is a need for reliable and accurate nondestructive timber assessment and allocation methods to evaluate standing and felled trees.	*Problem*: Reliable and accurate nondestructive timber assessment and allocation methods are needed to evaluate standing and felled trees.

Table 15. (Continued)

Research station— associated work unit, and plan components	Research work unit title, and period and activities covered		
	Activities: This research will evaluate and develop computer simulations, expert systems, and vision-system supported computer-aided hardwood sawmill edging and trimming. *Problem*: There is a need to identify, evaluate, and develop new or improved products that use our abundant nonselect hardwood sawtimber resources. *Activities*: Emerging and potential primary hardwood products will be identified along with new production concepts. *Problem*: There is a need to develop and demonstrate systems to make and market short-length lumber and dimension for profitable use of low-grade hardwood sawtimber. *Activities*: Research will determine through simulation and actual testing whether short lumber grading systems can be designed to facilitate the use of short logs, low quality logs, thick shortslabs, and/or low-grade lumber to make short lumber.	*Activities*: This research will investigate the use and application of ground-based digital imagery, combined with image processing software and product allocation software to estimate tree volume, product allocation, and tree grade for inventory purposes. *Problem*: There is a need to improve wood processing technology and equipment to efficiently evaluate and process wood resources. *Activities*: This research will develop and use expert systems and vision systems to support computer-aided and automated hardwood sawmill edging and trimming. *Problem*: There is a need to extend the life of resources, effectively refurbishing and reusing wood pallets and other solid wood products, or converting them to alternative high-value products.	*Activities*: This research will investigate the use and application of ground-based digital data collection, combined with information extraction software and product allocation software to estimate tree volume, product allocation, and tree grade for inventory purposes. *Problem*: Wood processing technology and equipment must be improved to efficiently evaluate and process wood resources. *Activities*: Develop and use expert systems and vision systems to support computer-aided and automated hardwood sawmill edging and trimming. *Problem*: There is a need to extend the life of resources, effectively refurbishing and reusing wood pallets and other solid wood products, or converting them to alternative high-value products.

Table 15. (Continued)

Research station— associated work unit, and plan components	Research work unit title, and period and activities covered		
		Activities: This research will investigate and evaluate current and new repairs for damaged and used wood pallets and compare these to new pallets.	*Activities*: This research will investigate and evaluate current and new repairs for damaged and used wood pallets and compare these to new pallets.
4703	Title unknown, 1994-99	Forest Operations Research to Achieve Sustainable Management, 1999-2005	
Area of Research applicability	No information available.	National	
Mission	To provide engineering knowledge and improved, economically viable forest operations for sustained resource management and develop an understanding of the interactions between biological and engineering systems in forest ecosystems.	To provide the science and technology integrating ecological and engineering disciplines to achieve economically and ecologically viable forest operations which are necessary for sustainable and socially acceptable forest resource management.	
Research problem and activities	*Problem*: There is a need to sufficiently understand the implications and interactions between biological systems, soil, and hydrologic site factors, and forest operations for the design and development of environmentally sound forest operations.	*Problem*: There is a need to develop a better understanding of the effect of forest operations systems on the ecological processes of forest ecosystems.	

Table 15. (Continued)

Research station— associated work unit, and plan components	Research work unit title, and period and activities covered		
	Activities: The cause-effect relationship between selected engineering variables and soil physical property changes will be established. Such information will be used to improve systems for forest operations. *Problem*: Current techniques and technologies are not always feasible or cost effective under a holistic approach to forest resource management. *Activities*: Approaches to solving this problem involve improving system design, management, and performance, or developing alternative concepts such as innovative technologies and small-scale machinery. *Problem*: The capabilities and limitations of the human element are not adequately considered in current forest operations, resulting in degraded safety, productivity, and system performance.	*Activities*: To understand more completely the complex machine-soil interaction in terms of temporal, spatial, vegetative, site, and climatic differences, and to integrate the results into a comprehensive systems approach to management for sustained above- and below-ground productivity. *Problem*: There is a need to develop new and innovative methods and technologies to reduce the ecological impacts of forest operations. *Activities*: Considering the impact roads have on the forest ecosystem, more work needs to be undertaken to design road systems which are acceptable based on the goal of sustainable forestry practice. *Problem*: Better information about the performance, cost, and operational ranges of new and existing forest operations systems is needed.	

Table 15. (Continued)

Research station— associated work unit, and plan components	Research work unit title, and period and activities covered		
	Activities: Three distinct research elements will be pursued—industrial safety management, effective personal protective equipment, and application of ergonomic principles to equipment design. *Problem*: Available analytical methods for forest operations are not sufficient to support the implementation of multi-resource, multi-scale management in southern forests. *Activities*: Research will focus on the development of a specific model for tactical planning of access and management activities.	*Activities*: Performance will be measured in terms of factors such as production, energy consumption, labor input, and capitalization. *Problem*: New approaches are needed to acquire and utilize information for planning and management of forest operations across landscapes. *Activities*: Studies will be conducted to develop tools for optimizing application efficiency of equipment on given sites, identifying those factors that most influence effectiveness of operations, and techniques to adapt systems to local conditions.	

Source: Forest Service documents.

In: Wood Utilization
Editor: J. E. Carter, pp. 109-129
ISBN: 1-60021-459-2

APPENDIX III: CSREES WOOD UTILIZATION RESEARCH CENTERS, FISCAL YEARS 1995-2005

This appendix presents information on CSREES' wood utilization research centers, including some of their objectives, specialty areas, and research activities over 11 years—fiscal years 1995 through 2005.

UNIVERSITY OF ALASKA WOOD UTILIZATION RESEARCH CENTER

This center specializes in assisting the Alaska Forest Products industry through research, extension, and education activities.

INLAND NORTHWEST FOREST PRODUCTS RESEARCH CONSORTIUM

The consortium uses a multidisciplinary, multi-institutional approach to solving forest operations and wood utilization problems unique to the Inland Northwest region. The consortium consists of the universities of Idaho and of Montana, and Washington State University.

Table 16. University of Alaska Wood Utilization Research Center—Activities Conducted under Grants for Wood Utilization Research and Product Development, Fiscal Years 2000-2005

Fiscal Year	Objectives	Approach/research activities
2000	Implement a broad-based program that will help Alaska become competitive in the value-added wood products industry by providing specific technical, business, and marketing assistance; develop a facility where promising new projects can be developed and tested; study the educational and training needs of sawmill operators (and others) involved in present and evolving wood products industries and then to respond with a client-centered approach to technical transfer and product development; make assistance available in processing and performance, marketing research, and development of wood and forest-based products.	• Conduct an overall needs assessment of educational and industrial clientele in Alaska's forest products industry • Use study results to identify and develop a variety of techniques that will be the basis for another research project to ascertain the most effective means of transferring technology to help entrepreneurs and their employees develop and apply technology for peak efficiency • Support ongoing efforts to develop in-grade testing specific to Alaska species
2001	Help Alaska's forest products industry develop new, cost-effective methods for converting wood and other tree and plant components to consumer products; define new markets; and, assist individuals and organizations by providing required training and research.	• Provide course offerings and technical training that are required by a dynamic industry • Support applied research projects that focus on the basic physical and mechanical properties of wood and other tree and plant material • Conduct research that identifies new products, technology, production processes, and markets; and, extension services to assist entrepreneurs and firms in the industry
2002	Help Alaska's forest products industry develop new, cost-effective methods for conversion of wood and other forest material to marketable products; emphasize forest products education identification of markets, and new and improved production system to create high-value products from low-value material; help Alaska become competitive in the value-added forest products industry by providing specific technical, business, and marketing assistance and a facility for developing and testing promising new projects.	Continues 2001 approach.

Table 16. (Continued)

Fiscal Year	Objectives	Approach/research activities
2003	Help overcome lack of representation on boards, committees, and organizations that create grading rules or standards and collect marketing information, which often results in barriers to the economic development and maximum use (technically and economically) of the products from the region. Conduct projects that assist with the restructuring of the forest products industry in Alaska; emphasize projects that have the potential to help rebuild the forest products industry in southeast Alaska.	• Continue support for an in-grade testing program at the Ketchikan Wood Technology Centers • Use samples in accordance with ASTM D-143 to evaluate the properties of small clear samples of western hemlock lumber, focusing on the relationship between specific gravity and the evaluated properties • Use results as the basis for recommendations for proceeding with an Alaska Wood Density Survey to allow development of new small clear strength values for western hemlock harvested in Alaska
2004	Conduct research programs to help restructure the forest products industry in Alaska; design research projects to respond to the needs of the industry and people in Alaska; develop projects based on needs identified by ongoing research and studies, direct request from industry organizations, educational institutions, and government agencies.	• Continue an ongoing University of Alaska project at the Ketchikan Wood Technology Center to review the basic properties of Alaska species • Develop updated strength values for Alaska species, based on testing of small clear samples
2005	Continue testing Alaska wood species for unique properties for value-added products; report the chemical determination of Alaska birch bark by species and region using the bark from harvested trees; conduct an integrated study that inventories and maps harvestable nontimber forest products, plant species, and potential harvest areas; document traditional and current uses of special forest products in four additional southeast Alaska native communities; make recommendations for marketing efforts for value-added Alaskan wood and special forest products.	• Use standard tests to determine the potential for using Alaskan birch in value-added industries • Study plant species and potential harvest areas to inventory nontimber forest products • Continue to define consumer reaction to terms used to describe lumber produced from Alaska species and to promote value-added wood products made from Alaskan wood and special forest products

Source: GAO's analysis of CSREES data.

University of Maine Wood Utilization Research Center

This center specializes in all aspects of utilization concerning species indigenous to the New England area.

Table 17. Inland Northwest Forest Products Research Consortium—Activities Conducted under Grants for Wood Utilization Research and Product Development, Fiscal Years 1999-2005

Fiscal year	Objectives	Approach/research activities
1999	Conduct forest products research in the inland northwest region of Idaho, Montana, and Washington on the species quantity, and quality of raw material; the capabilities and processes needed by the forest products industry to convert this resource to wood products; and the potential for wood products from alternate species designed for specific applications.	• Investigate regional problems in harvesting, processing and potential products from the wood resource expected in the future by joint research teams at the Forest Products Department, University of Idaho; Bureau of Business and Economic Research, University of Montana; and Wood Materials and Engineering Laboratory, Washington State University • Conduct research projects in resource assessment, manufacturing, and harvesting processes; raw material properties; and new and value-added wood products
2000	Continues 1999 objectives.	Continues 1999 activities.
2001	Continues 1999 objectives.	Continues 1999 activities.
2002	Continues 1999 objectives.	Continues 1999 activities.
2003	Continues 1999 objectives.	Continues 1999 activities.
2004	Conduct research to help make regional sawmills more efficient, assess the color modification of wood via ohmic heating, evaluate wood thermoplastic composites for bridge decking, develop flexible wood composite sheets for sound or thermo insulation, analyze the region's wood products industry, develop seismic design parameters for log shear wall, and evaluate thermoplastic binders derived from lignin.	The Inland Northwest Forest Products Research Consortium represents a cooperative effort between the Forest Products Department of the University of Idaho, the Bureau of Business and Economic Research at the University of Montana, and the Wood Materials and Engineering Laboratory at Washington State University. The Consortium takes an interdisciplinary, multi institutional approach to solving forest operations and utilization problems unique to the Inland Northwest, with an emphasis on those associated with new forest management regimes.

Table 17. (Continued)

2005	Conduct research to (1) improve the physical and visual properties of ponderosa pine lumber sawn from small-diameter trees; (2) optimize the color of wood via ohmic heating; (3) harden and dimensionally stabilize wood; (4) characterize the Inland-Northwest wood products industry; (5) enhance the utilization of small-diameter timber; (6) develop a fully wood-based wood composite; (7) develop seismic design parameters for log shear walls; and (8) develop alternative treating methods for oriented stand composites.	Continues 2004 activities.

Source: GAO's analysis of CSREES data.

Table 18. University of Maine Wood Utilization Research Center—Activities Conducted under Grants for Wood Utilization Research and Product Development, Fiscal Years 1995-2005

Fiscal year	Objectives	Approach/research activities
1995	Increase knowledge of the properties of timber species in New England to improve the efficiencies and environmental compatibility of existing industries, and to develop new products that could help expand the forest products industrial base.	• Biodegradation and preservation of wood • Computer-aided processing of northeastern species • Fundamental properties of northeastern species • Structural engineering • Wood-based composites for structural applications • Wood chemistry
1996	Continues 1995 objectives.	Continues many 1995 activities, and adds the following:
1997	Continues 1995 objectives.	• Improved environmental compatibility of pulping technologies • Chlorine-free sulfur-free high-yield pulps from northeastern hardwood species • Wood fiber composites utilizing paper deinking sludge • Feasibility of glulam beams reinforced with fiber-reinforced plastic sheets • Oxidation degradation of lignocellulose by low molecular weight chelators isolated from brown-rot fungi • Novel technology for the detection of wood-degrading fungi

Table 18. (Continued)

Fiscal year	Objectives	Approach/research activities
1998	Continues 1995 objectives.	• Improve design to increase efficiency of alkaline oxygen pulping • Technical feasibility and cost-effectiveness of combining fiber- reinforced polymers • Identify the role of redox chemical reactions in the breakdown of lignocellulosic materials through X-ray absorption and fine structure spectroscopy
1998	Continues 1995 objectives.	• Experimental techniques of fluorescence, genetic, and biochemical analysis to understand the physiology of wood decay fungi • Effect of steaming and dry heat on the resonant frequency and frequency response of wood • Forecast the service life of laminated composites comprised of glass fibers bonded with catalyzed PVAc to red pine or red maple species
1999	Continues 1995 objectives.	• Determine major factors affecting the price of wood for use in pulp manufacture and softwood dimension manufacture • Evaluate the effects of precommercial thinning on wood properties of spruce-fir forests • Predict warp potential in spruce-fir studs using ultrasonic waves
2000	Continues 1995 objectives.	• Develop a wood-concrete connection and evaluate its behavior under static and fatigue load • Identify the causes and costs of idle logging product capacity, wood supply, labor, and other issues related to Maine logging • Evaluate dry matching of 2x4 SPF studs from green condition using three drying schedules and variation in moisture content loss along the length of the dimension samples, measure the warp, and compare restricted loss and unrestricted loss samples
2001	Evaluate the basic processing and feasibility of manufacturing structural composite lumber from northeastern wood species using a long-strand, high yield, log breakdown procedure; assess the commercial feasibility of stock glulam beams produced from Maine hardwood resources; assess the relation of local differences in shrinkage for the level of warp in red and white pine lumber.	• Restore a log reducer to operating condition and use it to crush small-diameter logs; after resin addition, use an RF press to produce structural scale material to determine mechanical properties • Merchandize 3,000 board feet of #3 common 4/4 hardwood by length and width to established yields for e-rated lamination stock • Harvest red and white pine trees and saw lumber from two stands. Obtain shrinkage rate samples from the logs. Measure and correlate full sized lumber shrinkage with sample rates.

Table 18. (Continued)

Fiscal year	Objectives	Approach/research activities
2002	Develop an optimized preassembled narrow shear wall system using advanced OSB; improve the strength and stiffness properties of polyolefin wood composites by modifying the post-die process conditions; investigate the material requirements planning, logistics, and conversion efficiency of northeastern pulp mills.	• Conduct static and cyclic connector tests using screws and ring shank nails and develop allowable design values for walls using ICBO AC 130 criteria • Monitor cooling rates via thermocouple and physical and mechanical properties • Obtain information from publicly available data sources supplemented by on-site interviews with key personnel involved in inventory planning and control at both groundwood and kraft pulp mills
2003	Study impact of processing additives on extruder operating parameters and properties of polypropylene wood-plastic composites; quantify several factors that may affect the levels of VOC release from hardwoods and softwoods; evaluate the competitive position of the Maine paper industry.	• Identify the influences of maleation, zinc stearate, EBS wax, fire retardant, neustrene, and HALS on the mechanical performance of extruded wood polypropylene composites through ASTM mechanical and weathering tests • Measure VOCs using established procedures on fresh logs of red oak, white oak, hard maple, and cherry; perform other tests with white pine that is untreated or dipped in various fungicides • Monitor both primary and secondary resources to meet the objectives; use interviews and databases for each resource
2004	Improve the technical performance and efficiency of products that use wood as the main constituent.	• Conduct design parameters, panel, and connection design through finite element modeling and experimental testing of full-scale panel systems • Use a variety of commercial additives to prepare extruded wood-plastic materials using commercial scale equipment available in the laboratory • Obtain specimens of several Maine wood species from sawmills; obtain composite panels from regional manufacturers and measure specific heat capacity using a differential calorimeter
2005	Discover underlying science and develop technology that will both improve the ability to provide sustainable forest-based products for societal needs and help increase the global competitiveness of the U.S. forest products industry.	• Use enzymatic pretreatment of pulps to evaluate commercial laccases and catechol dioxgenases prior to oxygen delignification to reduce the environmental footprint • Elucidate fundamental physiological and biochemical pathways of fungi to develop strategies of biological techniques to control microbial biodegradation and preserve wood and wood products • Evaluate the use of chemical and physical foaming agents in extrusion systems to reduce the density of wood plastic composites and the application of near InfraRed (near-IR) technologies to wood plastics composites production to provide on-line product quality information

Source: GAO analysis of CSREES data.

Michigan State University Wood Utilization Research Center

The center specializes in sustainable hardwood utilization, with a focus on wood preservation, wood composite materials, and genetic engineering of necessary wood properties for specific product development.

Table 19. Michigan State University Wood Utilization Research Center—Activities Conducted under Grants for Wood Utilization Research and Product Development, Fiscal Years 1995-2005

Fiscal year	Objective	Approach/research activities
1995	Improve hardwood lumber yield by examining alternatives to current rough-mill practices for saw log conversion and evaluate short log and underutilized species for veneer production for concealed furniture parts; establish standards or evaluating composite surface quality with comparison of contact and noncontact methods; continue hardwood preservation research.	• Improvement of the hardwood saw log conversion process • The surface quality and stability of wood and wood products • Preservative treatment of hardwood • Recycling of treated lumber, untreated recycled wood fiber, and wood ash • Laser-cutting of wood and wood composites
1996	Improve hardwood lumber yield and utilization; explore recycling opportunities to reduce wood consumption; improve quality of wood composites; compare log yields by grade sawn and live sawn; evaluate economics of short log and underused species for veneer production; set standards for evaluating composite surface quality; complete comparison of stability model with lab-produced OSB.	• Hardwood preservation project • Complete evaluation of particleboard and fiberboard made with recycled newsprint • Make and evaluate composites made from recycled treated and demolition wood • Explore new pulsed laser cutting approach • Economic analysis of log improvement study and recycling projects
1997	Improve yield and value from hardwood logs through conversion process improvements; extend forest resources by preserving wood and composites and by recycling treated wood; use mixed eastern hardwoods to manufacture particleboard; rationalize internal bond and thickness OSB swell test specimen; promote forest sustainability and certification of eastern hardwood forests.	• Create real log database for computer-simulated sawing comparisons • Determine mechanism of CCA treatment in soft maple using logs harvested at different seasons • Determine properties of composites made of recycled treated wood and conduct durability tests • Compare properties of single-species and mixed species furnish particleboards • Expose hardboard siding and other hardboard materials to cyclic swelling and shrinking to establish secondary stability

Table 19. (Continued)

Fiscal year	Objective	Approach/research activities
1998	Extend hardwood forest resources by preserving wood and wood composites and by recycling treated wood; improve stability of wood composites; rationalize internal bond and thickness of OSB swell test specimen; recycle wood ash from power plants; promote forest sustainability and certification of eastern hardwoods.	• Determine species of copper in wood after various treatments and relationship to effectiveness of preservative • Develop a method of removing CCA from treated wood for recycling of fibers • Make and test particleboards of single and known blend of hardwood species • Monitor surface energy changes of treated and untreated wood, glued and not glued and relate changes to properties of wood composites
1999	Extend forest resource by preserving wood and composites and by recycling treated wood; evaluate the low performance of CCA-treated hardwood; evaluate mixed hardwood species growing in Michigan as raw material for particleboard and OSB manufacture.	• Compare properties of single species and mixed species furnished • Evaluate horizontal density distribution, among other things, of representative commercial OSB samples and determine the influence of specimen size on measurements
2000	Extend the service life of forest products, particularly hardwood species, by using preservatives, reusing treated wood, recycling wood removed from service, and applying biotechnological means for producing high decay resistant wood; evaluate the recovery of CCA from treated wood removed from service; evaluate mixed hardwood species growing in Michigan as raw materials for particleboard and OSB manufacture.	• Conduct field and laboratory tests to determine the toxic threshold and the residual chemicals of several commercial wood preservative formulations used to treat silver maple, beech, red oak, elm, willow, and hard maple after 5 years' exposure • Manufacture particleboard and OSB single species and mixed species furnished of hardwoods growing in Michigan and evaluate their properties evaluated to determine the effect of species furnish
2001	Extend the service life of forest resources, particularly hardwood species, by using preservatives, reusing treated wood, recycling wood removed from service, applying biotechnological means to produce high-decay resistant wood; investigate environmental benign chemicals used in crop protection as wood preservatives; evaluate mixed hardwood growing in Michigan as a source of raw materials for flake board and OSB.	• Use laboratory soil-block tests and field tests to evaluate the biological performance of some environmentally benign wood preservatives • Test wood cement-bonded particleboard and wood fiber thermoplastic made using recycled materials to determine their physical, mechanical, and biological performance according to well-established standards • Measure and use surface energy of raw materials to explain the adhesion phenomenon between wood fibers and the matrix • Compare the properties of OSB made with single and mixed species furnished • Use genetic manipulation technology to increase the levels of wood extractives that render them decay resistant

Table 19. (Continued)

2002	Increase use of hardwood species for exterior applications: use environmentally benign chemicals; reuse and recycle wood products from demolition as raw materials for wood composites; develop processes to use sawdust from the furniture industry as raw materials to manufacture commercially viable and value-added products; investigate effects of major economic and social forces on timber supply and demand in subregions of the Lake States; and develop biotechnology to produce value-added wood products.	• Screen chemicals with low-to-negligible impact on the environment as potential preservatives • Use a continuous extrusion process to manufacture wood plastic composites • Analyze the relationship between sawlog price, labor demand, and capital investment in the Lake States' hardwood sawmill industry • Use a metabolic engineering approach to produce value-added hardwood products
2003	Continues 2002 objectives.	Continues 2002 activities.
2004	Continues 2002 objectives.	Continues some 2002 activities and adds genomics of decay resistance and wood growth will be studied using micro array analysis to determine the genes that are involved in the commercially important biological processes
2005	Increase the use of hardwood species for exterior applications: use environmentally benign chemicals as wood preservatives and reuse and recycle wood products as raw materials for wood composites; develop processes to use sawdust and wood fiber to produce commercially viable extruded wood-plastic composites; use biotechnology to evaluate the control of genes that express important wood characteristics and to develop new value-added products.	• Screen chemicals with low impact on the environment as wood preservative • Use laboratory soil block tests and field exposure tests to determine the biological effectiveness and study samples to determine the degree of fixation in the wood • Use continuous extrusion and co-extrusion processes to manufacture wood-plastic composites as well as particleboards without using formaldehyde-based adhesive

Source: GAO's analysis of CSREES data.

University of Minnesota, Duluth, Wood Utilization Research Center

This center specializes in helping existing small and mid-size wood products companies remain or become competitive by conducting research and development and forest productivity work in hybrid poplar, red pine, and other Minnesota species; and by forming new regional industries based on forest products materials and technologies. This is done to ensure the sustainability of the forest products industry

Table 20. University of Minnesota, Duluth, Wood Utilization Research Center—Activities Conducted under Grants for Wood Utilization Research and Product Development, Fiscal Years 1995-2005

Fiscal year	Objective	Approach/research activities
1995	Develop private sector forest products opportunities in the Upper Midwest, particularly Minnesota; help existing small and mid-size forest products companies implement technologies that can help economic growth.	• Supply technical assistance • Help with business plans and marketing • Assist with cost accounting and inventory systems • Perform prototype manufacturing, and conduct product or process research and development Much of the work will involve the transfer of technologies previously developed at the center or developed at other research institutions.
1996	Continues 1995 objectives.	Continues 1995 activities.
1997	Continues 1995 objectives.	Continues 1995 activities.
1998	Continues 1995 objectives.	Continues 1995 activities.
1999	Continues 1995 objectives.	Continues 1995 activities.
2000	Continues 1995 objectives.	Continues 1995 activities.
2001	Help the sustainable development of private sector forest products opportunities in Minnesota and the Upper Midwest; emphasize helping existing small and mid-size wood products companies remain or become competitive through research and development or specific technical or business expertise; direct work in biotechnology and chemical extractives to form new industries in the region, benefiting farmers, small landowners, and large regional forest products companies; undertake promising independent projects that may lead to new job creation without an identified industry partner.	Generally continues 1995 activities and adds the following: • Undertakes product or process research and development within program areas—biotechnology, chemical derivatives, biobased composites and manufacturing efficiency, wood materials and engineering, and secondary wood products
2002	Continues most of 2001 objectives and adds specific subprojects in the areas of wood engineering, secondary wood products manufacturing, and biobased composites.	• Serve more than 50 companies as clients and strategic partners • Use funding from previous special grants to work with more than 100 different companies, resulting in the formation of new companies, and facilitating the rapid growth of others

Table 20. (Continued)

Fiscal year	Objective	Approach/research activities
2003	Help the sustainable development of private sector forest products opportunities in Minnesota and the Upper Midwest.	• Chemical extractives • Wood materials and engineering • Secondary wood products • Biobased composites • Manufacturing efficiency
2004	Assist with the sustainable development of private sector forest products opportunities in Minnesota and the Upper Midwest; help small and mid-size wood products companies remain or become competitive through research and development; form new regional industries.	• Lean manufacturing • Nondestructive evaluation • Portable wood finishing systems • Membrane press technology • Ready-to-assemble wood frame housing internal connectors • Lightweight sandwich panel • Phosphate bonded composites • ThermoWood • Larch tree extractives
2005	Assist with the sustainable development of private sector forest products opportunities in Minnesota and the Upper Midwest; help small and mid-size wood products companies remain or become competitive through research and development; form new regional industries based on forest products materials and technologies.	• Lean manufacturing concepts • Nondestructive evaluation technologies • Membrane press technology • Breathable and waterproof mineral-bonded strandboard • Iron phosphate impregnated wood products • Radio frequency identification • Process technology for aspen, poplar, and sycamore tree Extractives • Utilization of forest harvest residues for renewable energy

Source: GAO's analysis of CSREES data.

Mississippi State University Wood Utilization Research Center

This center specializes in timber harvesting, transportation, and economics; lumber manufacturing and processing; wood-based composite materials; protection and preservation of wood; wood chemistry; economic evaluation; and technology transfer.

Table 21. Mississippi State University Wood Utilization Research Center—Activities Conducted under Grants for Wood Utilization Research and Product Development, Fiscal Years 1995-2005

Fiscal year	Objectives	Approach/research activities
1995	Administer a continuing research program on using southern pines; strengthen existing efforts in wood utilization; provide support for research initiatives in these areas.	• Harvesting, transportation, and primary processing • Economic evaluation and technology transfer • Structural engineering • Wood chemistry • Protection and preservation of wood • Timber manufacturing and processing • Wood-based composite materials
1996	Continues 1995 objectives and adds timber harvesting.	Continues 1995 activities.
1997	Continues 1996 objectives.	Continues 1995 activities.
1998	Continues 1996 objectives.	Activities in five of seven broad program areas: • Harvesting, transportation, and primary processing • Economic evaluation and technology transfer • Wood chemistry • Protection and preservation of wood • Timber manufacturing and processing
1999	Administer a continuing program of research and technical assistance on utilization of southern pines, to strengthen existing efforts in wood utilization, and to provide support for new research initiatives in these areas.	Research to be conducted within five of the seven broad areas: • Harvesting, transportation and timber processing • Economic evaluation and technology transfer • Structural engineering • Protection and preservation of wood • Timber manufacturing and processing
2000	Administer a continuing program of research and technical assistance to improve the use and value of southern timber resources; strengthen existing efforts in timber harvesting and wood utilization; and support new research initiatives in these areas.	Continues 1999 activities.
2001	Continues 2000 objectives.	Research to be conducted within five of the seven broad areas: • Harvesting and transportation • Economic and market evaluation and technology transfer • Wood engineering and wood-based composites • Wood protection and biodeterioration • Timber manufacturing and processing
2002	Continues 2000 objectives.	Continues 2001 activities, except for harvesting and transportation.

Table 21. (Continued)

Fiscal year	Objectives	Approach/research activities
2003[a]	Continues 2000 objectives.	• Forest resources • Economic and market evaluation and technology transfer • Engineered wood products • Wood protection and biodeterioration • Timber manufacturing and processing
2004	Continues 2000 objectives.	• Utilization of wood-based materials in housing • New manufacturing systems for wood-based industry • Fiber and chemicals from wood • Timber harvesting and wood utilization in Mississippi
2005	Continue 2000 objectives.	• Biotechnology and fundamental research • Chemicals and energy from wood • New manufacturing systems for wood-based industries • Performance of wood structures and housing

Source: GAO's analysis of CSREES data.

[a]In fiscal year 2003, the Mississippi center began receiving a committee-directed grant to conduct a statewide forest resources inventory. This grant was included in the center's fiscal year 2003 proposal; thereafter, the center has submitted separate proposals for the inventory.

North Carolina State University Wood Utilization Research Center

This center specializes in wood machining and tooling technology.

Table 22. North Carolina State University Wood Utilization Research Center—Activities Conducted under Grants for Wood Utilization Research and Product Development, Fiscal Years 1995-2005

Fiscal year	Objective	Approach/research activities
1995	Improve understanding of the interface between the woodworking machine and the wood product, including the machine power source, spindle assembly, cutting tool, method of attachment of the tool to the spindle, cutting tool geometry and cutting edge materials and treatment, and resulting work piece accuracy and surface finish quality.	· Machine and tool vibration and stability · Tool materials and tool wear mechanisms · Process monitoring and control · Surface measurement technology
1996	Continues 1995 objective.	Continues 1995 activities.
1997	Establish a national educational and research resource to foster improvements in woodworking machine and tool performance; focus on the machine-tool-work piece interface to develop a better understanding of spindle and tool dynamics, tool materials and tool wear, and machining conditions.	Adapt and develop technology to further understanding of the wood-machining process, including cross-transfer of computer models from metal working to woodworking to help relate tool forces to the machining conditions (process monitoring) and work piece surface quality
1998	Establish education and research programs to help the wood machining and tooling industries conserve natural resources, compete with foreign markets, and compete with other building material industries; provide industry with graduates knowledgeable in wood machining practices and the means to educate employees in the latest technology through technology transfer. Conduct applied research focusing on the machine-tool-work piece interface.	Activities conform with objectives.
1999	Continues 1998 objectives.	Objectives and descriptions of research activities are similar to objectives.
2000	Continues 1998 objectives.	• Adapt existing technologies from other disciplines, such as the metal-cutting and aerospace industries • Use high speed machining techniques to achieve higher throughputs • Adapt process monitoring techniques developed for metal cutting to wood machining • Use surface quality evaluation techniques to monitor the machining process as well as the condition of the work piece.

Table 22. (Continued)

Fiscal year	Objective	Approach/research activities
2001	Foster research and education to achieve an improved understanding of tool/work piece interaction phenomena, including enhancing wood utilization and wood products manufacturing efficiency through increased tool life, improving surface quality, reducing machine and tool maintenance problems, improving cutting accuracy, and increasing machine productivity.	Continues 2000 activities, and adds Conduct abrasive machining research, resulting in significant • energy savings
2002	Continues 2001 objectives.	Continues 2001 activities.
2003	Continues 1998 objectives.	Continues 2001 activities.
2004	Help the wood machining and tooling industries conserve natural resources, compete with foreign markets, and compete with other building material industries; provide industry with graduates knowledgeable in wood machining practices and with the means to educate employees in the latest technology through technology transfer efforts; conduct applied research focusing on the machine-tool-work piece interface.	Continues 2001 activities.
2005	Help the wood machining and tooling industries conserve natural resources, compete with foreign markets, compete with other building material industries, as well as provide these industries with graduates knowledgeable in wood machining practices. Focus on the machine-tool-work piece interface to better understand and improve tool materials, machine and tool designs, and process monitoring and control techniques.	Continues some 2001 activities.

Source: GAO's analysis of CSREES data.

Oregon State University Wood Utilization Research Center

This center specializes in science, technology, and business practices that will enhance the domestic and global competitiveness of the U.S. wood products industry, especially in the western United States; this will ensure more efficient use of available wood resources. A special emphasis is placed on training future scientists, researchers, and practitioners.

Table 23. Oregon State University Wood Utilization Research Center—Activities Conducted under Grants for Wood Utilization Research and Product Development, Fiscal Years 1995-2005

Fiscal year	Objective	Approach/research activities
1995	Meet environmental goals during timber harvest and forest product manufacture, leading to sustained timber production; extend the forest resource base through improved manufacturing and processing, developing new wood composites, and extending the service life of wood; develop new structural applications for wood; exploit wood extractives as alternatives to current preservatives, pesticides and adhesives, and as a potential source of pharmaceuticals.	• Applied and basic studies on forest harvesting and other operations such as road-building • Basic studies on wood and other materials properties to use in applied research on composites manufacture and testing • Basic and applied studies in wood engineering, wood preservation, and wood chemistry
1996	Continues 1995 objectives.	Continues 1995 activities.
1997	Continues 1995 objectives.	Continues 1995 activities.
1998	Continues 1995 objectives.	Continues 1995 activities.
1999	Continues 1995 objectives.	Continues 1995 activities.
2000	Meet environmental goals, particularly water quality, during timber harvest and forest products manufacture, leading to sustained timber production; improve understanding of slope stability in forested environments; extend the forest resource base through improved timber harvest and road-related practices, manufacturing and processing, developing new wood products, and extending the life of wood; develop new structural applications for wood; explore new ways to detect wood defects; and understand the effects of different silvicultural practices on wood quality.	• Applied and basic research studies, including modeling, on forest harvesting and other operations, such as road-building and maintenance • Basic research on slope stability in forest environments • Basic studies on properties of wood and other materials to use in applied research on composites manufacturing and testing; basic and applied studies in wood engineering, wood preservation, and wood chemistry
2001	Develop the relationships between temperature, humidity, airflow, and the levels of VOC emissions from wood during the kiln drying process; create a decision support system that helps forest planners select efficient transportation routes for montane forest operations; develop a model to determine the life-cycle assessment of structural wood products, considering the steps in manufacturing.	• Dry wood under a variety of conditions, and panel products manufactured to compare emissions at both dryer and press—especially to look for any effects of drying conditions on subsequent press emissions • Use global information systems and economic analysis techniques to analyze terrain and road construction and maintenance costs and to generate optimization techniques • Use an existing computer model to develop life-cycle data for a selected number of wood-based composite products

Table 23. (Continued)

Fiscal year	Objective	Approach/research activities
2002	Develop the knowledge and technology necessary to balance the sustainable use of the nation's forest resources with the need to maintain a vigorous, competitive, domestic forest products industry.	• Aggregate existing optimization and simulation software modules into a unified system with a user-friendly interface and demonstrate effectiveness in solving typical manufacturing problems • Experimentally determine if slash bundling technology can be effectively used in harvesting of 1- to 4-inch-diameter trees in high-fire risk stands • Select three promising "smart sensor" technologies and experimentally evaluate for field efficacy, accuracy, repeatability, and cost in a forestry application
2003	Continues 2002 objectives.	• Collect data from selected ongoing commercial logging operations and use regression analysis to develop forecasting tools • Use traditional market research and telephone surveys after in-depth interviews of 16 "bellwether firms" • Modify a base stochastic drying model for recent research on presorting and other features
2004	Develop the science, technology, management approaches, and business practices that will enhance the domestic and global competitiveness of the U.S. wood products industry, especially in the Pacific Northwest.	• Use a mix of log production control gaming in field evaluations and design harvest experiments • Evaluate bioactivity of species-specific mill residues and analyze chemical constituents of those that test positive • Experimentally assess the effect of range of decay fungi on selected composite material properties
2005	Continues 2004 objectives.	• Employ new life-cycle inventory and assessment models to benchmark current manufacturing practices and analyze process alternatives • Physically measure hygro-mechanical and other properties using compression tests and controlled changes in climate conditions • Apply field, laboratory, and computer-simulation techniques to a series of 5-10 ha model stands

Source: GAO's analysis CSREES data.

UNIVERSITY OF TENNESSEE WOOD UTILIZATION RESEARCH CENTER

This center specializes in southern Appalachian hardwood utilization and manufacturing of composite materials.

Table 24. University of Tennessee Wood Utilization Research Center—Activities Conducted under Grants for Wood Utilization Research and Product Development, Fiscal Years 1999-2005

Fiscal year	Objective	Approach/research activities
1999	Apply statistical process control methodologies to manufacturing hardwood lumber for improved lumber thickness control; develop dimensionally stable wood-based composites panel products for improved performance and durability; characterize the hardwood sawmill in Tennessee to ensure future productivity and competitiveness.	• Have researchers from the Tennessee Forest Products Center work together to attain the program objectives • Conduct research in cooperating sawmills, in the laboratory, and by mail survey, of sawmill owners • Analyze data and prepare reports and publications • Make presentations to industry groups, professional societies, and research cooperators
2000	The Tennessee Quality Wood Products Initiative will address understanding sources of variation in manufacture of hardwood flooring and cabinets; the Tennessee Hardwood Lumber Processing Initiative will address efficiency in manufacture of hardwood lumber; an international technology transfer conference on dimensional stability of wood-based composites will bring together researchers from around the world to address this subject through oral presentations and poster papers.	Continues 1999 approach and adds Research in flooring and cabinet plants and secondary wood • products manufacturers in Tennessee • International technology transfer conference on wood-based composites
2001	Apply statistical process control to improve hardwood lumber processing; evaluate microwave technology for wood drying; and develop technology to improve OSB panel with emphasis on thickness swell optimization.	• Center researchers work with other university and industry partners to attain program objectives • Conduct research in cooperating industries and the laboratory, and by conference training • Analyze data and prepare reports and publications • Make presentations to industry groups, professional societies, and research cooperators

Table 24. (Continued)

Fiscal year	Objective	Approach/research activities
2002	Improve the efficient and effective use of the nation's hardwood resources by expanding their use in composite wood products like medium-density fiberboard and OSB. Two research projects focus on minimizing raw material loss by improving process monitoring and control methods. The work will explore the information available from new spectroscopic sensors while developing new statistical tools for data mining and information presentation.	• Center researchers work with other university, government, and industry partners to accomplish the program objectives • Conduct research primarily in the Center's laboratories • Compile and analyze experimental data for presentation in reports and scientific publications, and make presentations to industry groups, professional societies, and cooperating researchers
2003	Define the effect of molecular level orientation in regenerated cellulose fibers on the properties of carbon fibers made from this starting material; develop dynamic mechanical analysis to assess mechanical properties of red oak at elevated temperature and moisture contents; determine the effect of adhesive resin content and cure level on strength characteristics of wood fibers.	• Build on recent reprogramming of Center's research to include a focus on renewable composites and develop low-modulus, carbon fibers from cellulose fibers regenerated from NMMO solution • Apply nanoindentation to study fundamental questions of interphase structure and properties to composite performance and use this tool to generate new information on the material properties of the wood cell wall
2004	Focus on delivering new insight into the performance of wood in heterogeneous materials to better define the fundamental parameters that dictate product performance; better understand the role of the wood/polymer interface in composites, and develop improved systems for monitoring and control of the manufacturing process.	• Develop new algorithms to model product properties based on input from multiple sensors, including novel near infrared fingerprints • Use microscopy and thermal analysis to study species effects on wood-plastic composite structure and properties • Develop spectroscopic imaging to characterize resin penetration and chemistry in veneer-composite bondlines
2005	Focus on developing new information on processing and performance of extruded wood-polymer composite materials; better understand the role of the wood and polymer interface in composites, better define interfacial structure-property relationships, and develop improved systems for monitoring and control of the extrusion process; develop novel wireless sensors for moisture content monitoring during the drying process.	• Develop new approaches to monitor and model extruded composite properties based on input from multiple sensors, including novel near infrared fingerprints • Conduct research on wood-polymer composites to explore the effect of copolymer architecture on interfacial structure and adhesion with amorphous polymers • Use dynamic mechanical analysis and spectroscopic methods to assess characteristics of the wood-polymer interphase

Source: GAO analysis of CSREES data.

UNIVERSITY OF WEST VIRGINIA WOOD UTILIZATION RESEARCH CENTER

This center specializes in improving the utilization of upland hardwoods in Appalachian forests.

Table 25. University of West Virginia Wood Utilization Research Center—Activities Conducted Under Grants for Wood Utilization Research and Product Development, Fiscal Years 2004-2005

Fiscal year	Objective	Approach/research activities
2004	Provide research leadership to the forest products sector in the Appalachian region and increase the success of the wood products industry: assess the fundamental properties of Appalachian hardwoods relative to improved use and manufacturing; improve opportunities for value-added products developed from residues, poor quality resources, and underutilized upland hardwood species; and enhance the global competitiveness of Appalachian hardwood industries by promoting their quality and efficient use.	• Condition assessment of logs using ground-penetrating radar (GPR): use GPR to develop nondestructive scanning technology for identifying subsurface defects in hardwood logs • Oak logging residues: determine lumber and component yields of low-quality logs and bolts remaining after timber harvesting in West Virginia • Determine economic feasibility of converting logging residues into value-added products for the lumber, tie, pallet, and component industries • Increase use of low-quality wood: increase use of oak by investigating knife angles and projections during stranding for OSB panel production
2005	Provide research leadership to the forest products sector in the Appalachian region and increase the success of the wood products industry.	Continues 2004 activities and adds • Value recovery through merchandizing hardwood log products • Enhancement of commercial competitiveness through application of advanced technologies

Source: GAO analysis of CSREES data.

In: Wood Utilization
Editor: J. E. Carter, pp. 131-137
ISBN: 1-60021-459-2

Appendix IV: Budget Authority for the Forest Service's Research Work Units and for the CSREES Wood Utilization Research Centers

This appendix presents budget authority information for the Forest Service, information on FTE scientists and support staff for the Forest Service, and budget authority information for CSREES wood utilization research centers, from fiscal years 1995 through 2005.

Table 26. The Forest Service's Wood Utilization Research Work Units' Budget Authority, Fiscal Years 1995-2005

Nominal dollars in thousands											
Work units	1995	1996	1997	1998	1999	2000	2001	2002	2003	2004	2005
Forest Products Laboratory											
4502	$549	$351	$317	$655	$660	$660	$643	$680	$722	$806	$673
4701	333	333	333	474	478	478	468	469	553	532	425
4703	1,338	1,125	1,150	976	983	983	960	962	987	1,107	1,107
4706	874	894	894	1,104	1,113	1,113	1,086	1,163	1,188	1,265	1,265
4707	610	271	280	523	527	527	515	591	616	697	697
4708	1,031	1,031	1,031	[a]	[a]	[a]	[a]	[a]	[a]	[a]	[a]
4709	1,395	1,195	1,395	1,340	1,350	1,350	1,316	1,319	1,389	1,877	1,877
4710	1,147	1,267	1,267	2,042	2,058	2,058	2,206	2,211	2,236	2,231	2,231
4712	1,333	1,228	1,363	1,349	1,359	1,359	1,427	1,430	1,476	1,474	1,474
4714	893	953	953	965	973	973	1,091	1,093	1,118	1,193	2,445
4715	870	910	910	[a]	[a]	[a]	[a]	[a]	[a]	[a]	[a]
4716	1,917	1,502	1,502	1,419	1,430	1,430	2,182	2,514	2,461	2,555	875
4718	739	335	[a]	[a]	[a]	[a]	[a]	[a]	[a]	[a]	[a]
4719	1,495	1,300	1,300	1,335	1,345	1,345	1,228	1,231	1,327	1,308	1,308
4722	918	933	933	972	980	980	956	958	1,028	1,013	1,013
4723	1,249	1,249	1,249	1,354	1,365	1,365	1,294	1,372	1,397	1,394	925
4724	[a]	[a]	[a]	[a]	[a]	[a]	629	630	655	638	638
4725	[a]	[a]	[a]	[a]	[a]	[a]	[a]	[a]	[a]	[a]	775
4851	905	923	923	1,218	1,229	1,229	1,923	1,928	1,935	1,935	1,485
Subtotal—Forest Products Laboratory	$17,596	$15,800	$15,800	$15,726	$15,850	$15,850	$17,924	$18,551	$19,088	$20,025	$19, 213
Northeastern Research Station											
4701	1,114	1,059	1,459	1,459	1,059	1,059	1,139	1,137	1,131	1,117	1,084
4751	300	200	200	200	200	261	233	235	246	250	248
4803	857	400	478	650	1,050	1,030	1,159	1,157	1,150	1,136	1,103

Table 26. (Continued)

Nominal dollars in thousands											
Work units	1995	1996	1997	1998	1999	2000	2001	2002	2003	2004	2005
4805	112	108	129	126	126	145	155	155	167	168	161
Pacific Northwest Research Station											
4865	1,442	1,131	1,277	1,227	1,569	2,423	2,637	2,671	2,638	2,717	2,644
Pacific Southwest Research Station											
4202							225	225	225	206	164
Rocky Mountain Research Station											
4156				200	300	300	300	300	300	300	300
Southern Research Station											
4104	569	475	375	330	330	336	368	374	381	391	401
4701	1,089	1,100	1,100	1,100	1,110	1,171	1,226	1,226	1,226	1,250	1,183
4702	448	400	398	398	402	420	475	495	495	491	479
4703	200	200	200	200	200	200	200	200	200	200	200
Total	$23,727	$20,873	$21,416	$21,616	$22,196	$23,195	$26,041	$26,726	$27,246	$28,251	$27,179

Source: Forest Service data.

[a]Indicates the research work unit was not established, was discontinued, or was not doing any wood utilization research in the given year.

Table 27. FTE Scientists and Support Staff in the Forest Service's Wood Utilization Research Work Units, Fiscal Years 1995-2005

Work unit	Staff	1995	1996	1997	1998	1999	2000	2001	2002	2003	2004	2005
Forest Products Laboratory	Total scientists	67.2	69.0	64.3	60.6	56.4	55.8	59.0	58.5	57.2	62.8	59.6
	Total support	78.0	73.5	61.5	60.4	61.9	60.3	62.3	61.5	62.0	58.3	57.3
	Total FTE	145.2	142.5	125.8	121.0	118.3	116.1	121.3	120.0	119.2	121.1	116.9
4502	Scientists	4.2	4.2	2.8	2.3	2.1	2.2	2.2	2.2	2.2	2.2	2.1
	Support	1.1	1	2	2	2	2	2	2	2.2	2	1.1
4701	Scientists	2	2	3	2	2	1	1	2	2	2	1
	Support	0	0.2	1.9	1.9	1.9	1.9	1.9	1.9	1.9	1.9	1.9
4703	Scientists	4	4	4	4	4	2.6	3.3	3	3	3	3
Work unit	Staff	1995	1996	1997	1998	1999	2000	2001	2002	2003	2004	2005
	Support	3.4	3.3	3.5	3.5	3.5	3.5	3.5	3.5	3.5	4.4	4.5
4706	Scientists	2	2	4.5	5	4	4	3	3	3	5	6
	Support	4.7	6	6	6	6	6	6	6	6	9.4	9.5
4707	Scientists	1.2	1	2.1	2	2	2	2	2	2	2	2
	Support	2.5	2.5	2.5	2.5	2.5	1.5	1.5	1.5	1.5	2.5	2.5
4708	Scientists	3.2	5.1	a	a	a	a	a	a	a	a	a
	Support	7	4.5	a	a	a	a	a	a	a	a	a
4709	Scientists	6	6	6.1	6	5	5.8	6	6	4.4	4	5
	Support	3.5	8	3.5	3.5	3.5	3.5	3.5	3.5	3.5	3.5	3.5
4710	Scientists	4	4	6	6	6	6	6.5	5	4.6	4.5	6
	Support	11	8.5	8.5	8.5	12	12	13.5	13.5	14.7	9.3	6.5
4712	Scientists	5	5	3.5	3	3	4	4	4	4	4	4
	Support	10	8.1	8.1	7	7	7	7	7	7	6.5	6.5
4714	Scientists	4	4	4	3	3	2.8	3.5	3	3	7.8	7
	Support	2.3	2	2	2	2	2	2	2	2.5	2.2	2.2
4715	Scientists	4	4	a	a	a	a	a	a	a	a	a
	Support	0.5	0.5	a	a	a	a	a	a	a	a	a

Table 27. (Continued)

Work unit	Staff	1995	1996	1997	1998	1999	2000	2001	2002	2003	2004	2005
4716	Scientists	8.3	8.7	7	7	7	7	5.2	6	6.7	7	2
	Support	6.5	5.5	4	4	3	3	2.5	2.5	1.1	0.5	1.7
4718	Scientists	2	2	a	a	a	a	a	a	a	a	a
	Support	3.5	3.5	a	a	a	a	a	a	a	a	a
4719	Scientists	4	4	5	5	4	4	4	4	3.5	3.5	3
	Support	5.5	3.5	2.5	2.5	2.5	1.9	1.9	1.9	1.9	1.9	1.9
4722	Scientists	5	5	5	5	5	5	5	5	5	4.4	5
	Support	2.5	2	5	5	5	5	5	5	5	5	5
4723	Scientists	2	2	5	4	3	4	4	4	4	2	3
	Support	7.3	7.3	6	6	6	6	6	6	6	3.8	4.3
4724	Scientists	a	a	a	a	a	a	3	3	3	3	3
	Support	a	a	a	a	a	a	2	2	2	2	2
4725	Scientists	a	a	a	a	a	a	a	a	a	2	2.1
	Support	a	a	a	a	a	a	a	a	a	2.6	2.5
4851	Scientists	6.3	6	6.3	6.3	6.3	5.4	6.3	6.3	6.8	6.4	5.4
	Support	6.7	7.1	6	6	5	5	4	3.2	3.2	0.8	1.7
Northeastern Research Station	Total scientists	13	14	12	10	12	12	11	10	9	9	9
	Total support	10	9	9	10	8	10	10	11	11	11	10
4701	Scientists	7	8	7	6	5	5	5	4	3	3	3
	Support	4	4	4	4	3	5	5	6	6	6	6
4751	Scientists	1	1	1	1	1	2	1	1	1	1	1
	Support	1	1	1	1	1	1	1	1	1	1	1
4803	Scientists	4	4	3	2	5	4	4	4	4	4	4
	Support	5	4	4	5	4	4	4	4	4	4	3
4805	Scientists	1	1	1	1	1	1	1	1	1	1	1
	Support	0	0	0	0	0	0	0	0	0	0	0
Pacific Northwest Research Station— 4865	Scientists	7.5	5.8	5.8	5.8	5.8	8.8	8.8	7.8	10.1	9	9.3
	Support	8	6	5	5	5.5	7.5	7.5	7.5	6	4.5	7

Table 27. (Continued)

Work unit	Staff	1995	1996	1997	1998	1999	2000	2001	2002	2003	2004	2005
Pacific Southwest Research Station—4202	Scientists	[a]	[a]	[a]	[a]	[a]	[a]	1	1	1	1	1
	Support	[a]	[a]	[a]	[a]	[a]	[a]	0	0	0	0	2
Rocky Mountain Research Station— 4156	Scientists	[a]	[a]	[a]	0.1	0.1	0.1	0.1	0.1	0.1	0.1	0.1
	Support	[a]	[a]	[a]	0.0	0.0	0.0	0.0	0.0	0.0	0.0	0.0
Southern Research Station	Total scientists	10.5	9.5	8.5	8.5	8.5	7.5	6.7	6.7	8.7	8.7	8.7
	Total support	13.0	13.0	13.0	12.0	11.0	10.0	10.0	9.3	9.3	9.3	9.8
4104	Scientists	4	3	2	2	2	1	1.2	1.2	1.2	1.2	1.2
	Support	4	4	4	3	3	3	3	3	3	3	3
4701	Scientists	4	4	4	4	4	3	2	2	4	4	4
	Support	7	7	7	7	6	5	5	4.3	4.3	4.3	4.8
4702	Scientists	2	2	2	2	2	3	3	3	3	3	3
	Support	1.5	1.5	1.5	1.5	1.5	1.5	1.5	1.5	1.5	1.5	1.5
4703	Scientists	0.5	0.5	0.5	0.5	0.5	0.5	0.5	0.5	0.5	0.5	0.5
	Support	0.5	0.5	0.5	0.5	0.5	0.5	0.5	0.5	0.5	0.5	0.5
Forest Service	Total scientists	98.2	98.3	90.6	85.0	82.8	84.2	86.6	84.1	86.1	90.6	87.7
	Total support	109	101.5	88.5	87.4	86.4	87.8	89.8	89.3	88.3	83.1	86.1
	Total FTEs	207.2	199.8	179.1	172.4	169.2	172	176.4	173.4	174.4	173.7	173.8

Source: GAO's analysis of Forest Service data.

[a]Indicates the research work unit was not yet established, was discontinued, or was not doing any wood utilization research in the given year.

Table 28. Budget Authority for the CSREES Wood Utilization Research Centers, Fiscal Years 1995-2005

Nominal dollars in thousands											
Wood utilization research centers	1995	1996	1997	1998	1999	2000	2001	2002	2003	2004	2005
Alaska	[a]	[a]	[a]	[a]	[a]	$540	$596	$596	$585	$543	$602
Consortium	[b]	[b]	[b]	[b]	$514	457	514	504	511	460	509
Maine	$752	$752	$704	$704	824	732	824	807	807	736	717
Michigan	752	752	704	704	824	732	824	807	807	736	717
Minnesota	233	233	218	218	255	227	255	250	246	228	222
Mississippi[c]	752	752	704	704	824	732	824	807	1,260	1,154	1,180
North Carolina	289	290	271	271	317	282	317	311	306	283	276
Oregon	752	752	704	704	824	732	824	807	795	736	717
Tennessee	[b]	[b]	[b]	[b]	421	374	421	412	412	376	417
West Virginia	[d]	[d]	[d]	[d]	[d]	[d]	[d]	[d]	[d]	418	463
Total	$3,530	$3,532	$3,305	$3,305	$4,805	$4,805	$5,400	$5,304	$5,730	$5,670	$5,820

Sources: Wood utilization research centers' annual special grant proposals.

Note: Totals may not add due to rounding.

[a] This center was added in fiscal year 2000.

[b] This center was added in fiscal year 1999.

[c] In fiscal year 2003, the Mississippi center began receiving a committee-directed special grant to inventory forest resources statewide. This grant was included in the center's fiscal year 2003 proposal, and thereafter the center has submitted separate proposals for the inventory, which were included in the center's total budget for fiscal years 2004 and 2005.

[d] This center was added in fiscal year 2004.

In: Wood Utilization
Editor: J. E. Carter, pp. 139-141

ISBN: 1-60021-459-2

APPENDIX V: GAO CONTACT AND STAFF ACKNOWLEDGMENTS

GAO CONTACT

Robin M. Nazzaro (202) 512-3841

STAFF ACKNOWLEDGMENTS

In addition to the contact named above, Andrea Wamstad Brown, Jacqueline Adams Cook, Richard Johnson, Rebecca Shea, Jay Cherlow, Carol Herrnstadt Shulman, Jeremy Ames, and Jaelith Hall-Rivera, made key contributions to this report.

GAO'S MISSION

The Government Accountability Office, the audit, evaluation and investigative arm of Congress, exists to support Congress in meeting its constitutional responsibilities and to help improve the performance and accountability of the federal government for the American people. GAO examines the use of public funds; evaluates federal programs and policies; and provides analyses, recommendations, and other assistance to help Congress make informed oversight, policy, and funding decisions. GAO's commitment to good government is reflected in its core values of accountability, integrity, and reliability.

Obtaining Copies of GAO Reports and Testimony

The fastest and easiest way to obtain copies of GAO documents at no cost is through GAO's Web site (www.gao.gov). Each weekday, GAO posts newly released reports, testimony, and correspondence on its Web site. To have GAO e-mail you a list of newly posted products every afternoon, go to www.gao.gov and select "Subscribe to Updates."

Order by Mail or Phone

The first copy of each printed report is free. Additional copies are $2 each. A check or money order should be made out to the Superintendent of Documents. GAO also accepts VISA and Mastercard. Orders for 100 or more copies mailed to a single address are discounted 25 percent. Orders should be sent to:

U.S. Government Accountability Office 441 G Street NW, Room LM Washington, D.C. 20548

To order by Phone: Voice: (202) 512-6000 TDD: (202) 512-2537 Fax: (202) 512-6061

To Report Fraud, Waste, and Abuse in Federal Programs

Contact:

Web site: *www.gao.gov/fraudnet/fraudnet.htm E-mail: fraudnet@gao.gov*

Automated answering system: (800) 424-5454 or (202) 512-7470

Congressional Relations

Gloria Jarmon, Managing Director, JarmonG@gao.gov (202) 512-4400 U.S. Government Accountability Office, 441 G Street NW, Room 7125 Washington, D.C. 20548

Public Affairs

Paul Anderson, Managing Director, AndersonP1@gao.gov (202) 512-4800 U.S. Government Accountability Office, 441 G Street NW, Room 7149 Washington, D.C. 20548

INDEX

A

B

C

D

E

F

J

K

L

M

Q

R

S

T

U

V

W

Y

Z